THE MATHEMATICS OF GAMES
AND GAMBLING

NEW MATHEMATICAL LIBRARY

Published by

THE MATHEMATICAL ASSOCIATION OF AMERICA

The New Mathematical Library (NML) was begun in 1961 by
the School Mathematics Study Group to make available to
high school students short expository books on various topics
not usually covered in the high school syllabus. In a decade
the NML matured into a steadily growing series of some
twenty titles of interest not only to the originally intended
audience, but to college students and teachers at all levels.
Previously published by Random House and L. W. Singer, the
NML became a publication series of the Mathematical Association of America (MAA) in 1975. Under the auspices of the
MAA the NML will continue to grow and will remain dedi-
cated to its original and expanded purposes.

THE MATHEMATICS
OF GAMES
AND GAMBLING

by

Edward W. Packel

Lake Forest College

28

THE MATHEMATICAL ASSOCIATION
OF AMERICA

Library of Congress Catalog Card Number: 80-85037

Complete Set ISBN 0-88385-600-X

Vol. 28 0-88385-628-X

Manufactured in the United States of America

Note to the Reader

This book is one of a series written by professional mathematicians in order to make some important mathematical ideas interesting and understandable to a large audience of high school students and laymen. Most of the volumes in the *New Mathematical Library* cover topics not usually included in the high school curriculum; they vary in difficulty, and, even within a single book, some parts require a greater degree of concentration than others. Thus, while the reader needs little technical knowledge to understand most of these books, he will have to make an intellectual effort.

If the reader has so far encountered mathematics only in the classroom work, he should keep in mind that a book on mathematics cannot be read quickly. Nor must he expect to understand all parts of the book on first reading. He should feel free to skip complicated parts and return to them later; often an argument will be clarified by a subsequent remark. On the other hand, sections containing thoroughly familiar material may be read very quickly.

The best way to learn mathematics is to *do* mathematics. The reader is urged to acquire the habit of reading with paper and pencil in hand; in this way mathematics will become increasingly meaningful to him.

The authors and editorial committee are interested in reactions to the books in this series, and hope that readers will write to: Anneli Lax, Editor, New Mathematical Library, NEW YORK UNIVERSITY, THE COURANT INSTITUTE OF MATHEMATICAL SCIENCES, 251 Mercer Street, New York, N. Y. 10012.

The Editors

NEW MATHEMATICAL LIBRARY

Other titles in preparation

Preface

The purpose of this book is to introduce and develop some of the important and beautiful elementary mathematics needed for rational analysis of various gambling and game activities. While the only formal mathematics background assumed is high school algebra, some enthusiasm for and facility with quantitative reasoning will also serve the reader well. The book will, I hope, be of interest to:

1) Bright high school students with a good mathematics background and an (often related) interest in games of chance.
2) Students in elementary probability theory courses who might appreciate an informal supplementary text focusing on applications to gambling and games.
3) Individuals with some background in mathematics who are interested in some common and uncommon elementary game-oriented applications and their analysis.
4) That subset of the numerate gambling and game-playing public who would like to examine the mathematics behind games they might enjoy and who would like to see mathematical justification for what constitutes "good" (rational) play in such games.

One guiding principle of the book is that no mathematics is introduced without specific examples and applications to motivate the theory. The mathematics developed ranges from the predictable concepts of probability, expectation, and binomial coefficients to some less well-known ideas of elementary game theory. A wide variety of standard games is considered along with a few more recently popular and unusual activities such as backgammon and state lotteries. Generally it is not the play of the game, but the reasoning behind the play that provides the emphasis.

Some readers may be temporarily slowed down or frustrated by the omission of detailed rules for a few of the games analyzed. There are two important reasons for such omissions. First, I believe that initial exposure to a new game should be acquired through experience rather than a formal study of rules. Thus, I would encourage readers unfamiliar with backgammon to find knowledgeable partners and to learn by playing. In a classroom setting, the promise of a class tournament is an excellent way to encourage students to master the rules before any mathematical analysis is engaged in. Secondly, the book is not intended to be a "how to" document and I would urge readers interested in detailed rules and neatly packaged advice for backgammon, bridge, poker, or horse racing to consult one of the many specialized books on the particular topic (see the Bibliography). I give little "advice" that is not justified by prior mathematical analysis. In most cases it would be unreasonable to claim that this book will directly improve one's gaming ability, though I would hope that the insights provided into rational analysis and play will have some spinoff in that direction. Exceptions to this disclaimer may be backgammon, which is treated in some detail, and the various gambling house games, for which it is made quite clear (blackjack being a possible exception) that the optimal strategy is abstinence.

After an historical and literary initial chapter, subsequent chapters (except for the last) each include from 10 to 12 carefully selected exercises. These game-related questions are not primarily intended as drill problems, but rather for readers genuinely interested in expansion of ideas treated in the text. Some of the later exercises for each chapter may be quite challenging and some questions are open-ended, with emphasis on analysis and explanation rather than numerical answers. Chapter 2 provides the foundation for most of what follows, and should be mastered. The first two sections of Chapter 4 are needed also for Chapter 5. Otherwise chapters are, with minor exceptions, independent of one another. Sections referring to particular games can be included or omitted according to taste. The Bibliography presents, among other things, full reference information for books and articles referred to in abbreviated form in the text.

No serious judgments on the morality of gambling are intended in the book, though readers will no doubt draw inferences from the tone and tenor of my remarks at various points. While it should be clear that I am enthusiastic about games of chance and skill, I hope it is equally clear that I am even more enthusiastic about the mathematics behind such games. It is this latter enthusiasm that I hope will be transferred to the reader.

Numerous institutions and individuals have helped me in shaping this book. Lake Forest College and three separate groups of students took their chances with me over the last 3 years in an experimental freshman level course from which the final text has evolved. The Division of Social Sciences and Humanities at California Institute of Technology provided office space and technical assistance. Members of the New Mathematical Library subcommittee of the Mathematical Association of America have provided exceedingly thorough and helpful comments on the manuscript. I would also like to thank Bruce Cain, Charles Maland, and Kathryn Rindskopf for their valuable suggestions at various stages in the writing process. Finally, I am indebted to Gertrude Lewin for her typing of the first draft and Barbara Calli for her lovely artwork and typing of the final draft.

Contents

CHAPTER 1

The Phenomenon of Gambling

A selective history

Fascination with games of chance and speculation on the results of repeated random trials appear to be common to almost all societies, past and present. It is tempting to view prehistoric man's day-to-day existence as a continual series of gambles against nature with the ultimate stake, survival, as the nonnegotiable wager. With nature somewhat under control and a relatively predictable daily routine assured, man's necessity to gamble is relieved. Some of his newly found leisure is used to recapture, act out and celebrate those breathtaking earlier times.

The above is certainly a plausible scenario for the origins of games and gambling (and, for that matter, of art, poetry, politics, sports and war). While historical and archeological evidence does not currently exist to support fully claims that gambling is a primeval human instinct, the fact remains that gambling arose at a very early time and continued to survive and flourish despite legal and religious restrictions, social condemnation, and even very unfavorable house odds.

An early form of our six-faced die, found commonly on Assyrian and Sumerian archeological sites, is the astragalus (the bone just above the heel bone) of sheep, deer, and other animals of comparable size. Babylonian and early Egyptian sites (circa 3600 B.C.) provide clear evidence that polished and marked astragali were used along with colored pebbles (counters and markers) and a variety of game-type "boards." A suitably chosen astragalus will fall with four possible orientations, making it a natural gaming device or randomizer. The fact that the orientations do not occur with equal likelihood and that each astragalus has different randomizing characteristics would have discouraged any general theory and analysis of its behavior. This has also been cited by some (and discounted by others) as one reason why even a primitive body of ideas on probability did not emerge in ancient times.

Inevitably, the astragalus gave way to increasingly true versions of the modern six-faced die, but not without a number of offspring such as throwing sticks, other regular and irregular polyhedral dice, and various forms of purposely loaded and unfairly marked dice. By the time of the birth of Christ man finds himself well endowed with randomizers, board games, and the will and imagination to design and play an endless variety of additional games. As F. N. David states in her book *Games, Gods, and Gambling,*

> The idea of counting and enumeration is firmly established but not the concept of number as we know it now. The paraphernalia of chance events has been organized for man's pleasure and entertainment. Randomization, the blind goddess, fate, fortune, call it what you will, is an accepted part of life.

Playing cards appeared in Europe around the tenth century and their evolution has a colorful history. Especially interesting is the origin of the four suits and the royal succession of historical figures represented as particular jacks, queens, and kings. Having touched upon the beginnings of dice and cards, two staples of modern day gambling, we summarize in Table I data for other well known randomizing devices, some of the games played with them, and various other gambling activities common today. Since our overall interest is not simply in aspects of gambling but in the mathematics of practical and theoretical game situations, we also include information on games of pure skill (no randomizing factors involved) and on the theory of games.

With a variety of reasonably accurate randomizing devices and a newly emerging theory of probabilities to analyze them, gambling acquired new status in the seventeenth century. Indeed many of the finest scientific and philosophical minds of the times were excitedly engaged in discussing practical and theoretical problems posed by gaming situations. Cardano wrote *The Book on Games of Chance* in about 1520, though it was not published until 1663. Pascal and Fermat engaged in their famous correspondence on probability and gambling questions in 1654. In 1658 Pascal proposed his famous wager (see p. 23), which can be viewed as a game theoretic approach to the question of belief in God. The expectation concept was introduced by C. Huygens in 1657 in the first printed probability text, *Calculating in Games of Chance*. Throughout this time Leibniz made philosophical contributions to the foundations of probability. The remarkable development of probability theory in the latter half of the seventeenth century was

TABLE 1

Origins of Some Gambling-related Randomizers, Games, and Activities

Event	Date of origin	Region of origin
Astragalus	About 3600 B.C.	Middle East
Standard die	About 2000 B.C.	Egypt and elsewhere
Playing cards	10th century;	China;
	14th century	Western Europe
Roulette wheel	About 1800	France
Poker	About 1800	Louisiana territory
Backgammon	Ancient predecessors, 3000 B.C.	Middle East
	Modern rules, 1743	England
	Doubling cube, 1925	
Craps	Early 20th century	U.S.A., From English game of Hazard
Chuck-a-luck	Early 20th century	Traveling carnivals; from Hazard
Bingo and Keno type games	1880–1900	England; traveling carnivals
Bridge	Whist games, 16th century	England
	Auction, 1900's	England
	Contract, 1915–1929	U.S.A.
Lotteries	1st century;	Roman Empire;
	Middle Ages	Italy
Life Insurance	1583	England
Horse race betting	16th century	England
Chess	?B.C.–disputed	?
	7th century	India
Checkers	12th century	France
Go	About 1000 A.D.	China
Game theory	1928	Von Neumann, Germany
	1944	Von Neumann and Morgenstern, U.S.A.

climaxed by Jakob Bernoulli's *Ars Conjectandi* (written in the early 1690's, published in 1713), a brilliant forerunner of the theory, practice, and philosophical complexities which characterize the subject today. All of this scholarly attention and reasoned analysis might be thought to have stripped gambling of its aura of mystery, irrationality and notoriety, but the nature of the beast is not to be underestimated. Fortunes continued to be won and lost, bets were placed on both sides of games fair and unfair, and officialdom was as zealous as ever at restraining the wagers of sin.

The situation today has evolved in somewhat predictable fashion. People gamble on a vastly increased variety of games, most of whose optimal strategies and odds have been analyzed completely on paper or to a high degree of approximation on digital computers. Gambling is still controlled and in varying degrees illegal in most countries and frowned upon by most religions. Nonetheless, large and lucrative gambling meccas have sprung up in Nevada, Monte Carlo, Atlantic City, and increasingly many other centers throughout the world. States and countries which restrict private gambling by their inhabitants sponsor a dizzying variety of lotteries and take a healthy cut from the proceeds of racetracks. In virtually all phases of this organized gambling the odds are soundly stacked against the player; yet masses of people play happily, compulsively, poorly and well. On the local level bridge and backgammon clubs flourish, often with master points and pride rather than money at stake. Church groups sponsor bingo days and Las Vegas nights (for worthy causes) with consistent success. Private poker games, crap games, and illegal bookmaking are widespread. The elusive numbers racket in the United States has a turnover estimated at well over a billion dollars per year.

In summary, the phenomenon of gambling is ubiquitous, recognizing no geographic, social or intellectual boundaries. Its mystery and appeal are a sometimes random mixture of superstition, excitement, hope, escapism, greed, snobbery, and mathematical fascination. Gambling is a vital part of some lives and an important sidelight for many others. It is in some cases destructive and potentially addictive, and in others a delight and a diversion. Gambling is, in mild forms, an almost universal childhood activity; its play is characterized by extremes of rationality, rationalization and irrationality; it is a serious, large, and growing business. Gambling, with all of its diverse, paradoxical and fascinating qualities, is here to stay.

The gambler in fact and fiction

People who gamble can be subdivided into three possibly overlapping types. The casual gambler plays at stakes sufficiently small that maximum losses will not usually be a financial hardship. He of course hopes to win, but may not expect to win and brings a clear sense of enjoyment to the act of gambling. While losses may bring regrets, personal embarrassment, and even guilt and self-recrimination, these feelings pass quickly and are not dwelt upon. The casual gambler may not be aware of odds or strategic subtleties and may not even fully comprehend the rules of the game. He does, however, know how much time and money he can afford to spend, and is able to tear himself away when these limits are approached.

The compulsive gambler is happy primarily when in the act of gambling. The dizzying contrast of emotions between the exultation of winning and the despair of losing is fueled by the belief that the player is truly blessed with a forthcoming streak of luck or a self-discovered inexplicable system which must work. Consequently the size of bets will escalate whenever possible, and quitting either while ahead or behind is an act of supreme willpower or financial necessity. The ability to rationalize, to violate self-made promises, and to find a way to make the next bet is present in creative abundance. The parallel with alcoholism is apparent, and it was no longshot that a world wide organization entitled Gamblers Anonymous began in California in 1947.

What are the internal and external forces that drive a compulsive gambler to continue this often self-destructive behavior? A complete answer to this question, if indeed one exists, would take us too far afield here. We note, however, that psychological research over the past two decades provides some fascinating answers based upon experiments with animals and humans. Among the various reinforcement schemes used to induce repetitive behavior, the most effective have been the "variable ratio reinforcement schedules." Such schedules provide reinforcement in "random" fashion but with the frequency of reinforcement increasing with the number of repetitions. One can hardly imagine a random reinforcement schedule more effective than the payoffs on a roulette wheel or a slot machine!

The professional gambler may at times gamble casually or compulsively. In any case he gambles well and makes a good living out of it, for otherwise he would be in another profession. We are not referring here to racetrack owners, casino owners, numbers racketeers, and the like; for they, like bankers, do not work a gamble but a near sure thing. The professional gambler may have little mathematical background, but he fathoms odds, games, and people with seldom erring instinct. His professional activities involve private games such as poker and bridge, informed betting at racetracks, and well placed bets on any variety of games and events at odds over which he exercises careful scrutiny if not control. He may indulge as a whim in such casino games as Keno, roulette, craps, and slot machines, but only for pleasure or to feed the nonprofessional aspects of his gambling instinct. Indeed he realizes that nobody plays such house games regularly and profits from it. (A possible exception, as we shall see in Chapter 5, is blackjack or twenty one.)

Clearly the above are rather stereotyped descriptions of the three gambling types proposed. In fact, the casual gambler may have all the game sense of the professional or the passion of the compulsive player. The point is that, whatever the inclinations, the casual gambler plays occasionally, under control, and only with nonessential funds. The compulsive gambler may have an outwardly ordinary existence and regular employment. The compulsion may manifest itself at irregular intervals, but when it does the gambler is, in a sense, out of control and possibly out of a considerable sum of money. In the following paragraphs we attempt to add flesh to our gambling stereotypes by describing some specific characters from history and fiction.

One of the most remarkable yet unrecognized characters in history must be Girolamo Cardano. Born in Italy in 1501, his life, even by conservative accounts, is awe-inspiring in its fullness, vicissitude, notoriety, controversy and, above all, intellectual breadth and attainment. A brilliant medical student, Cardano became the most highly regarded and sought-after physician in Europe. He was one of the foremost scientific minds of his era and published numerous popular and scientific works. His book on games is judged to be a first and major step in the evolution of probability theory. He is a central figure in the famous mathematical controversy concerning priority in solving a class of third degree polynomial equations. His no doubt exaggerated but brutally frank and self-analytic *Autobiography* is the first of its kind and is still read by literature students today. Overall Cardano left 131 printed works and 111 additional books in manuscript (he claimed to have burned another 170).

In contrast to this impressive array of intellectual achievements, Cardano spent time with his family in the Milan poorhouse; developed bitter enemies (and loyal friends) throughout his life; engaged in constant controversy and debate on questions in medicine, science, and mathematics; and saw his oldest son executed for murder and his youngest son jailed and exiled as a thief. He dabbled in astrology, casting horoscopes for royalty and of Jesus Christ (after the fact). Partially as a result of this latter activity he was arrested and jailed as a heretic at age 69. Throughout all this Cardano gambled incessantly. A moving and rather supportive account of this stormy life can be found in Ore's book *Cardano, The Gambling Scholar*, which also includes a translation of Cardano's *The Book on Games of Chance*.

Cardano seems to have played and written about almost all types of games common in his time. He excelled in chess, a game accompanied then by much betting and handicapping. He played backgammon and other dice games along with a variety of card games including primero, an early version of poker. His moral observations on gambling are both perceptive and amusing, especially in view of his almost total inability to heed his own advice. Thus he says in *The Book on Games of Chance* in a section on "Who Should Play and When": "So, if a person be renowned for wisdom, or if he be dignified by a magistracy or any other civil honor or by a priesthood, it is all the worse for him to play." He follows later in this section with: "Your opponent should be of suitable station in life; you should play rarely and for short periods, in a suitable place, for small stakes, and on suitable occasions, or at a holiday banquet."

There is nothing "suitable" about Cardano's gambling. He seems to have gambled continually, for large stakes, and with all variety of men. He says in his *Autobiography*:

> From my youth I was immeasurably given to table games; through them I made the acquaintance of Francisco Sforza, Duke of Milan, and many friends among the nobles. But through the long years I devoted to them, nearly forty, it is not easy to tell how many of my possessions I have lost without compensation. But the dice treated me even worse, because I instructed my sons in the game and opened my house to gamblers. For this I have only feeble excuse: poor birth and the fact that I was not inept at the game.

Then, in a chapter on "Gambling and Dicing":

> In perhaps no respect can I be deemed worthy of praise, but whatever this praise be, it is certainly less than the blame I deserve for my immoderate devotion to table games and dice. During many years—for more than forty years at the chess boards and twenty-five years of gambling—I have played not off and on but, as I am ashamed to say, every day. Thereby I have lost esteem, my worldly goods, and my time. There is no corner of refuge for my defense, except if someone wishes to speak for me, it should be said that I did not love the game but abhorred the circumstances which made me play: lies, injustices and poverty, the insolence of some, the confusion in my life, the contempt, my sickly constitution and unmerited idleness, the latter caused by others. An indication of this is the fact that as soon as I was permitted to live a dignified life, I abandoned it all. It was not love for the game, nor a taste for luxury, but the odium of my position which drove me and made me seek its refuge.

In this poignant passage we see the gambling compulsion revealed in a way that shows how little some things have changed in four hundred years. Cardano was many things and a gambler on top of it all, and one wonders how he found the time and energy for everything, while feeling grateful that he did. For whatever the gambling urge did to his life, it has left us with a rich picture of a brilliant and possessed man plus a priceless first treatise on the mathematics of games and gambling.

A better documented and more specific picture of compulsive gambling can be seen in the life and writings of Fyodor Dostoyevsky (1821–1881). It is no doubt significant that the life of this great Russian author rivals that of Cardano in its turbulence and swings of fortune, but we focus here on the roulette playing episodes experienced and written about by Dostoyevsky. He first became seriously involved with roulette in August of 1863 in Wiesbaden, where he wrote: "I won 10,400 francs at first, took them home and shut them up in a bag and intended to leave Wiesbaden the next day without going back to the tables; but I got carried away and dropped half my winnings."[†]

In a letter to his sister-in-law requesting that his wife be sent some of the money he had won, he said of the roulette experience:

> Please don't think I am so pleased with myself for not losing that I am showing off when I say that I know the secret of how not to lose but win. I really do know the secret; it is terribly silly and simple and consists of keeping one's head the whole time, whatever the state of the game, and not getting excited. That is all, and it makes losing simply impossible and winning a certainty. But that is not the point; the point is whether, having grasped the secret, a man knows how to make use of it and is fit to do so. A man can be as wise as Solomon and have an iron character and still be carried away.... Therefore blessed are they who do not play, and regard the roulette wheel with loathing as the greatest of stupidities.

Dostoyevsky's lack of respect for the constancy of roulette expectation (see Chapter 2) is coupled with a belief that keeping his head and not being carried away are the key to sure winnings. Already we see the pattern of a man with a system engulfed by the gambling urge. A week after he sent the letter just quoted, Dostoyevsky wrote from Baden-Baden to his brother Misha to ask for a return of some of his winnings:

> My dear Misha, in Wiesbaden I invented a system, used it in actual play, and immediately won 10,000 francs. The next morning I got excited, abandoned the system and immediately lost. In the evening I returned to the system, observed it strictly, and quickly and without difficulty won back 3,000 francs. Tell me, after that how could I help being tempted, how could I fail to believe that I had only to follow my system strictly and luck would be with me? And I need the money, for myself, for you, for my wife, for writing my novel. Here tens of thousands are won in jest. Yes, I went with the idea of helping all of you and extricating myself from disaster. I believed in my system, too. Besides, when I arrived in Baden-Baden I went to the tables and in *a quarter of an hour* won 600 francs.

[†]Quotes from Dostoyevsky given here are from the Introduction and text of the Jessie Coulson translation of *The Gambler*.

This goaded me on. Suddenly I began to lose, could no longer keep my head and lost every farthing.... I took my last money and went to play; with four napoleons I won thirty-five in half an hour. This extraordinary luck tempted me, I risked the thirty-five and lost them all. After paying the landlady we were left with six napoleons d'or for the journey. In Geneva I pawned my watch....

So it goes in an irregular succession of roulette forays throughout Europe with recently acquired but sorely needed funds—requests for more funds, pawned watches, and unpaid hotel bills until 1871 when somehow Dostoyevsky gives up roulette. It is not surprising that the state of mind and the events which characterized the earliest of these experiences should appear in Dostoyevsky's beautiful short novel *The Gambler*, a remarkable profile of one kind of compulsive gambler. It is worth noting that the actual writing of this novel was itself a gamble, with free rights to all Dostoyevsky's past and future writings as the stake. Indeed, as a result of a variety of family and financial problems plus considerable procrastination, he found himself on October 4, 1865, with a deadline of November 1 for presenting his creditor with a work of at least 160 pages, with the above-mentioned stake as the penalty for failure. By dictating to a specially hired stenographer (later to become his second wife) Dostoyevsky completed *The Gambler* on October 31.

While it is often tempting but usually stretching a point to regard certain fictional characterizations as clearly autobiographical, there seems to be ample justification in the case of *The Gambler's* narrator and title character Alexis. Employed and traveling abroad as a tutor to the family of "the General", Alexis observes and interacts with the family as they holiday at Roulettenburg pretending to live in grand style, while counting on much coveted additional funds from the will of the very wealthy and ailing "Grandmama", rumors of whose death are in the air. Having become passionately and slavishly attached to the General's daughter Polina, Alexis is commissioned by her to win at roulette.

As did Dostoyevsky, Alexis introduces himself to roulette and has immediate positive reinforcement, leaving the table a dazed but excited winner. We already see in him several symptoms of the gambling compulsion: the fatalistic certainty that gambling will shape one's own destiny, the loss of control, the sensations of plummeting and soaring. But Alexis is not made truly aware of the depth of his compulsion until the arrival of Grandmama who, despite her illness, has herself carried to Roulettenburg in response to the General's telegrams of inquiry about her death. In her we see a portrait of a totally unlikely victim being energized and swept away by the idea of gambling. (One recalls stories in Las Vegas of a 75 year old man or an $8\frac{1}{2}$ months pregnant woman playing the slot machines or the roulette wheel 18 hours a day.)

Grandmama watches the wheel and becomes attached to the 35 to 1 payoff on zero. After a frantic series of escalating bets zero is again called, leaving Grandmama a triumphant winner. Alexis is possessed by other feelings:

> I was a gambler myself; I realized it at that moment. My arms and legs were trembling and my head throbbed. It was, of course, a rare happening for zero to come up three times out of some ten or so; but there was nothing particularly astonishing about it. I had myself seen zero turn up three times running two days before, and on that occasion one of the players, zealously recording all the coups on a piece of paper, had remarked aloud that no earlier than the previous day that same zero had come out exactly once in twenty-four hours.

As the perceptive reader might guess, the heavy winnings of her first outing (would she have escaped had zero stayed away?) lead Grandmama to gamble away her whole fortune in the next two days. Alexis refuses to be a part of Grandmama's downfall after her first day of losses and independently is compelled to attack the roulette tables in an attempt to restore the honor and peace of mind of Polina. With just one hour remaining before closing time he dashes to the tables, possessed of the necessity of winning heavily and quickly:

> Yes, sometimes the wildest notion, the most apparently impossible idea, takes such a firm hold of the mind that at length it is taken for something realizable.... More than that: if the idea coincides with a strong and passionate desire, it may sometimes be accepted as something predestined, inevitable, fore-ordained, something that cannot but exist or happen! Perhaps there is some reason for this, some combination of presentiments, some extraordinary exertion of will power, some self-intoxication of the imagination, or something else—I don't know: but on that evening (which I shall never forget as long as I live) something miraculous happened to me. Although it is completely capable of mathematical proof, nevertheless to this day it remains for me a miraculous happening. And why, why, was that certainty so strongly and deeply rooted in me, and from such a long time ago? I used, indeed, to think of it, I repeat, not as one event among others that might happen (and consequently might also not happen), but as something that could not possibly fail to happen!

Alexis knows, at least describing it in retrospect, that the experience for which he has been specially singled out by fate is upon him. In a spiralling sequence of bets he wins a fortune of two hundred thousand francs. The money is almost meaningless to Alexis, but the experience of winning it will never leave him. The reader is made convincingly aware that it is the act of gambling, the next crucial spin of the wheel, which will rule Alexis' life. The story closes with Alexis, having spent time in debtors prison and as a servant, in a poignant soliloquy as he weighs the choice between delivering himself to Polina (who is now wealthy and loves him) in Switzerland or taking his tiny stake and gambling anew.

The Gambler is much more than a narrative about gambling and gamblers, but we have focused on that brilliant and revealing aspect of the book. Hopefully this has given the reader increased feeling for the psychological and emotional dimension of the gambling phenomenon. With these "romantic" aspects in mind, we examine in the remaining chapters some of the more practical and rational aspects of games and gambling.

CHAPTER 2

Finite Probabilities and Great
Expectations

The probability concept and its origins

It is a significant, amusing and sometimes overlooked fact that modern-day probability theory, with its high degree of abstraction and increasingly widespread applicability, owes its origin almost entirely to questions of gambling. Having already mentioned the life and writing of Cardano, we now describe two gambling problems which are frequently cited as the real beginning of the science of probability.

The Chevalier de Méré was a gentleman gambler with considerable experience and, apparently, a good feel for the odds in dice. Having made considerable money over the years in betting with even odds on rolling at least one six in four rolls of a die, de Méré reasoned by a plausibility argument common in his time that betting on one or more double sixes in twenty four rolls of two dice should also be profitable. To his credit he noticed from hard experience that this latter wager was not doing well for him, and in 1654 he challenged his renowned friend Blaise Pascal to explain why. In a series of letters between Pascal and Pierre de Fermat, de Méré's difficulty was explained, and in the process the idea of probability, Pascal's famous triangle, and the ubiquitous binomial distribution emerged. We leave the computations showing that de Méré's double six bet was unwise to Exercise 2.2, but we warm up by showing now that his single six bet is mathematically advantageous.

Using a standard ploy in the calculus of probabilities, we turn the question around and ask for the probability of obtaining *no* sixes in four rolls. On *each* of the four rolls the probability of no six is 5/6. The various rolls are said to be *independent* if the outcome on each has no bearing on the outcome of any other roll. We shall soon formalize the fact that probabilities of independent events can be multiplied to give the probability of the occurrence of *all* the independent events. Since rolls of the dice are clearly independent, the probability of no six in all four rolls is

$$\left(\frac{5}{6}\right)\cdot\left(\frac{5}{6}\right)\cdot\left(\frac{5}{6}\right)\cdot\left(\frac{5}{6}\right) = \frac{5^4}{6^4} = \frac{625}{1296}.$$

Since we are interested in the opposite result, namely at least one six in four rolls, we conclude that the probability of this is

$$1 - \frac{625}{1296} = \frac{671}{1296},$$

a probability which exceeds 1/2. Thus the single six bet is a winner.

A second problem solved in the famous Pascal–Fermat correspondence is the "problem of points", a special case of which can be paraphrased in modern terminology as follows. Jimmy and Walter are playing a game which requires a player to score 5 points in order to win and each player has an equal chance of winning a given point. Jimmy is leading 4 points to 3 when the game is raided by the police. How shall Jimmy and Walter divide the stakes on the unfinished game (assuming the winner was to receive some fixed reward)? Clearly the "right" answer to this question depends on a variety of moral, sociological, and mathematical assumptions to be made by the solver, and it might be instructive to list an array of plausible answers and their rationales. It is the essence of a good probabilistic solution to a problem that under natural and clearly explained assumptions the answer is not only plausible, but demonstrably correct. We illustrate by solving this problem of points.

Our plan is to find Jimmy's probability of winning if the game had been completed and to divide the stakes in accordance with this probability. It is easy to see that if two more points were played a winner would have to emerge. Letting J denote a winning point for Jimmy and W a point for Walter, precisely four different outcomes (JJ, JW, WJ, and WW) can occur in these last two points. By assumption these four outcomes are *equally likely*.

Point 8	Point 9	Overall winner
J	J	J
J	W	J
W	J	J
W	W	W

Therefore by inspecting the above chart, we conclude that Jimmy has a probability of 3/4 of winning the game, and Walter a probability of

1/4. Hence the reward for winning should be split in a ratio of 3 to 1 in favor of Jimmy.

The above solution and the simple dice calculation which preceded it illustrate the clever yet simple nature of many probability calculations. There are frequently many possible arguments to choose from, but only those which obey the well-defined rules of probability theory will consistently yield correct answers. We proceed in what follows to illustrate some of these calculations, with the hope that they will make the intuitive idea of probability somewhat clearer. We then present working definitions for the probability of an event, after which we formalize some of the rules for calculating with probabilities.

Dice, cards, and probabilities

Consider a pair of dice, each of which is *honest*, by which we mean that each of the 6 faces has an equal probability (namely 1/6) of turning up. If the two dice are rolled simultaneously (or sequentially, for that matter), there are 36 different equally likely outcomes; they are listed below. To distinguish between the dice we assume that one is green and the other white, although the results are valid whatever pair of honest dice is used.

| Green | 1 1 1 1 1 1 | 2 2 2 2 2 2 | 3 3 3 3 3 3 | 4 4 4 4 4 4 | 5 5 5 5 5 5 | 6 6 6 6 6 6 |
| White | 1 2 3 4 5 6 | 1 2 3 4 5 6 | 1 2 3 4 5 6 | 1 2 3 4 5 6 | 1 2 3 4 5 6 | 1 2 3 4 5 6 |

It is important to realize that the appearance of a 6 and a 1 on the two dice can happen in two distinct ways (green 1, white 6 and green 6, white 1) and hence is twice as likely as the appearance of two 3's. With this idea and the knowledge that each of the 36 basic outcomes has probability 1/36, we can compute by simple counting the number of ways and the probability of rolling a given total with two dice. Thus there are five distinct ways of totaling 6 (as indicated above), and consequently a probability of 5/36 of obtaining this total. We give a complete reckoning below.

Total	2	3	4	5	6	7	8	9	10	11	12
Number of ways	1	2	3	4	5	6	5	4	3	2	1
Probability	1/36	1/18	1/12	1/9	5/36	1/6	5/36	1/9	1/12	1/18	1/36

There are many other questions we could ask about dice probabilities, some of which will be addressed in due time. For now we urge the reader to understand and acquire some facility with the above counting arguments, and we move on to some similar elementary calculations involving playing cards.

A standard poker or bridge deck has 52 cards, 13 of each suit. If we assume perfect shuffling and honest dealing, then each card in the deck is equally likely to appear. Thus as long as player A has no specific knowledge of any cards that have been dealt, his probability of being dealt a card of a given suit is $13/52 = 1/4$, his probability of being dealt a card of a given rank (like King or Three) is $4/52 = 1/13$, and his probability of being dealt a specific card (like the Queen of Spades) is $1/52$. The reasoning so far is much as in the case of a pair of dice, where there are 36 equally likely possible outcomes. One distinction, bothersome to many, is that the above probabilities are not altered by the fact that some of the cards may have already been dealt to other players *as long as they have not been exposed to player A.* The reader is urged to convince himself of this and of the related fact that it does not matter from a probability standpoint if the hands in a card game are dealt (with no dishonest motive!) out of order.

As cards do become exposed to a given player, an important difference from the dice situation emerges. Cards from the deck are *drawn without replacement*, so the fact that a Heart (or even a Club) has already appeared affects the probability that the second card seen will be a Heart. This is to be contrasted with successive rolls of a pair of honest dice, where the appearance of several consecutive double sixes has no bearing (theories of "hot" and "overdue" dice notwithstanding) on the probability of a double six on the next roll. Returning to our deck of cards, we hope the reader will be convinced after some thought and consideration of equally likely outcomes that:

a) The probability of being dealt a pair in two cards is $3/51$.
b) The probability of getting a Jack or a Diamond in a single draw is $16/52$.
c) If 9 cards have been seen by player A, one of which is a Seven, his probability of the next card being a Seven is $3/43$.
d)[†] The probability of being dealt 5 Spades (a Spade flush) in 5 cards is

$$\frac{13}{52} \cdot \frac{12}{51} \cdot \frac{11}{50} \cdot \frac{10}{49} \cdot \frac{9}{48} = .000495,$$

which is just a bit less than $1/2000$.

[†] Hint: getting a spade on a first card and getting a spade on a second card given that the first card is a spade are independent events. The probability of getting spades on the first two cards is the product $(13/52)\,(12/51)$ of the probabilities of these independent events. Continue this argument.

Roulette, probability, and odds

The previous examples should indicate that the probability of a given event is a number between 0 and 1 which provides a measure of the likelihood of its occurrence. In the cases we have seen so far, each event can be broken down into a number of equally likely *elementary events* (36 for a pair of dice and 52 for a full deck of cards). Probabilities of events can then be determined by careful counting as follows:

Probability of event $E =$

$$\frac{\text{Number of elementary events in } E}{\text{Number of possible equally likely elementary events}}.$$

There are many probabilistic situations which cannot be broken down in any meaningful way into equally likely elementary events (such as the event of a snowfall in San Francisco on Christmas day). An empirical approach to probability which accords with intuition for some types of repeatable events proposes

$$\text{Probability of event } E \approx \frac{\text{Number of successful occurrences of } E}{\text{Number of trials}},$$

where $\approx$ means "is approximately equal to", and the approximation becomes better and better as the number of trials increases.

The reader should note that we have just presented two different notions of probability. Both are intuitively reasonable and very important in analyzing and simulating probabilities in games of chance. The conceptual differences between them are at the heart of much debate and controversy in the philosophical foundations of probability theory. Fortunately, the two approaches are linked by a central result in probability theory which says, among other things, that in the limit as the number of trials approaches infinity the two definitions coincide. We do not consider these deep and important ideas further.

Once the probability $p(E)$ of an event E is defined, it is common practice to refer to the *odds for* E and the *odds against* E. These concepts are related by:

$$\text{Odds for } E = \frac{p(E)}{1 - p(E)},$$

where the odds are expressed (when possible) as a ratio of whole numbers. Thus if the probability of E is $1/4$, the

$$\text{odds for } E = \frac{1/4}{3/4} = \frac{1}{3}$$

and we use the notation $1:3$ for these odds (read "one to three"). As might be expected, the odds against E are the reciprocal of those for E; so in the example above, the odds against E are $3:1$. Working in the other direction, it is not hard to deduce that, if the odds for event E are $a:b$, then

$$p(E) = \frac{a}{a+b}.$$

Thus, for instance, if event E has odds $7:3$ *against*, it has odds $3:7$ *for* and a probability of $3/(3+7) = 3/10 = .3$.

All of this is further complicated (though any enthusiastic gambler absorbs it with ease) by the practice of telling the bettor the odds against an event rather than the odds for it. The numerator in the odds then reflects the amount of profit on a successful wager in the amount of the denominator. Thus a \$2 bet at $7:2$ odds (against) will, if successful, result in a \$7 profit (plus the return of the \$2 stake).

To clarify the above ideas we apply them to a Las Vegas roulette wheel (as opposed to the European wheel mentioned in Chapters 1 and 5). The Vegas wheel is divided into 38 congruent (and supposedly equally likely) sectors numbered 1 through 36, 0, and 00. The numbers 1 through 36 consist of 18 red numbers and 18 black (0 and 00 are green). Figure 1 shows the betting layout for such a wheel and illustrates some of the allowable bets. Shading on a number indicates that it corresponds to a black number on the wheel. Besides the bets indicated directly by the labels on the betting table, a number of other bets are possible as indicated by the solid circular chips drawn on the table. From top to bottom these bets signify:

- a bet on the single number zero (any other single number bet is also permissible).
- a bet on the adjacent (on the table) numbers 5 or 8 (2 ways to win).
- a bet on the 10, 11, 12 row (3 ways to win).
- a bet on the four numbers 13, 14, 16, 17.
- a bet on the six numbers 28 through 33.

Letting E denote the event that a spin of the wheel results in a 10, 11, or 12 and using our first notion of probability, we find that the

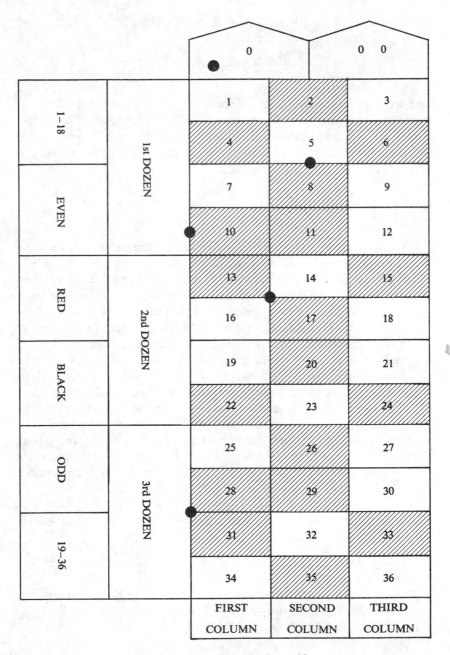

Figure 1.   A Las Vegas roulette table.

probability of $E = 3/38$. Thus we conclude that the odds for $E = 3:35$ and the odds against $E = 35:3$.

In Table 2 we present a more complete list of Las Vegas roulette bets, true odds against them, and the *house odds*. The house odds are all of the form $r:1$ where r thus represents the amount a player wins on a \$1 bet. It should be clear from comparing the true odds with the house odds why casino owners are very fond of the game of roulette (see Example 4, p. 23 for an elaboration on this theme).

TABLE 2

True and House Odds in Las Vegas Roulette

Type of bet[†]	True odds	House odds
Color (Red or Black)	20 : 18	1 : 1
Parity (Even or Odd)	20 : 18	1 : 1
18 #'s (1–18 or 19–36)	20 : 18	1 : 1
12 #'s (columns or dozens)	26 : 12	2 : 1
6 #'s (any 2 rows)	32 : 6	5 : 1
4 #'s (any 4 number square)	34 : 4	8 : 1
3 #'s (any row)	35 : 3	11 : 1
2 #'s (adjacent)	36 : 2	17 : 1
Single #'s	37 : 1	35 : 1

Compound probabilities: The rules of the game

Having looked at a variety of examples of computing probabilities of simple events by counting, we now present rules for computing probabilities of compound events. The following terminology will be needed:

The event *not A* occurs when event A fails to occur in a given *experiment* (rolling of dice, dealing of cards, spin of the wheel, etc.).

The event *A or B* occurs when, in a given experiment, A occurs or B occurs (or both). If it is impossible for A and B to occur simultaneously, they are said to be *disjoint* events.

The event *A then B* occurs when, in successive experiments which are *independent* (the results of the first experiment have no influence upon the results of the second), A occurs in the first experiment and B in the second. [Example: A = top card is spade; B = top card is red; each experiment deals the top card off a full 52 card deck.]

We now summarize the basic rules of probability theory (called axioms in more formal treatments) in Table 3. For an event E we let

[†]The symbol # stands for "number".

TABLE 3

Probability Rules and Illustrations

	Rule	Illustration
1.	For any event E $0 \leqslant p(E) \leqslant 1$	$0 \leqslant$ # successes $\leqslant$ # trials, so $0 \leqslant \dfrac{\text{\# successes}}{\text{trials}} \leqslant 1$
2.	$p(\text{impossible event}) = 0$ $p(\text{sure thing}) = 1$	$p(\text{rolling 13 on two dice}) = 0$ $p(\text{getting an Ace when 49 cards are dealt} \\ \text{from a bridge deck}) = 1$
3.	$p(\text{not } E) = 1 - p(E)$	$p(\text{at least one 6 in two dice})$ $= 1 - p(\text{no 6 in two dice})$ $= 1 - 25/36 = 11/36$
4.	If A and B are disjoint $p(A \text{ or } B) = p(A) + p(B)$	$A = $ draw a red Ace; $B = $ draw a black card $p(A \text{ or } B) = \dfrac{2}{52} + \dfrac{26}{52} = \dfrac{28}{52} = \dfrac{7}{13}$
5.	If A and B are events in independent successive experiments, $p(A \text{ then } B) = p(A) \cdot p(B)$	$A = $ draw an Ace or King $B = $ roll doubles with two dice $p(A \text{ then } B) = \dfrac{8}{52} \cdot \dfrac{1}{6} = \dfrac{1}{39}$

$p(E)$ denote the probability of event E occurring. Likewise $p(A$ then $B)$ denotes the probability of the occurrence of the compound event A then B.

The rules given above are an important part of the mathematics of games and gambling. Though they are easy to state and to work with in simple situations, their general application is more difficult and requires a clear understanding of the underlying events and what can be assumed about them. Indeed, the subtle application of these rules is a source of frustration as well as delight to many students of elementary probability theory. We will not rely heavily here on the elusive ability to reason creatively with probabilities, but we will employ the above rules in simple situations when called for, and the reader is urged to master their elementary use. In particular, the concepts of disjoint events and independent experiments should be understood. The exercises at the end of this chapter are intended to facilitate this mastery.

Mathematical expectation and its application

Perhaps the most important idea in making rational decisions in the face of uncertainty (a common task in gambling and many games) is the concept of mathematical *expectation*. In this section we will motivate this concept with a few examples, give it a general formulation, and then apply it to a variety of situations. One of our most significant results will be a negative one—that the *type* of bet placed in the Las Vegas version of roulette has no influence on expected gains or losses. (It will be seen later that the *size* of bet placed also has no effect upon a player's mathematical expectation per unit wagered.)

Expectation example 1:   You have the option to play a game in which two fair coins are flipped. Payoffs are as follows:

> both heads— you win $2
> both tails— you win $3
> one of each— you lose $4

Should you play and, if you do play, what is your expected gain or loss in an "average" game?

Solution: In an average play of 4 games you will obtain "both heads" once, "both tails" once, and "one of each" twice. In this 4 game sequence your net gain will be $1 \cdot \$2 + 1 \cdot \$3 + 2 \cdot (-\$4) = -\$3$—i.e., you will lose $3 over 4 games. Thus you should not play the game, but if you do you can expect to lose an average of $(1/4) \cdot \$3 = 75$ cents each time you play. To obtain this 75 cent expectation more directly, observe that $p(\text{both heads}) = 1/4$, $p(\text{both tails}) = 1/4$ and $p(\text{one of each}) = 1/2$. Hence

$$\tfrac{1}{4}(\$2) + \tfrac{1}{4}(\$3) + \tfrac{1}{2}(-\$4) = -\$.75$$

is obtained by multiplying each of the probabilities by the corresponding reward or payoff, with losses accompanied by negative signs.

Expectation example 2:   You roll a single die and the house pays you 12 coupons for rolling a six, 8 coupons for rolling an odd number, and nothing otherwise. How many coupons should you pay the house each time you roll in order to make the game a fair one?

Solution: Using the probability approach described in Example 1 above, we compute $p(\text{six}) = 1/6$, $p(\text{odd}) = 1/2$, $p(\text{other}) = 1/3$. In a single play your expectation is $\tfrac{1}{6}(12) + \tfrac{1}{2}(8) + \tfrac{1}{3}(0) = 6$ coupons. You should therefore pay 6 coupons at the outset of each roll to make the game *fair*—that is, an even proposition on the average. If you incorporate the 6 coupon price per roll into the payoff scheme and re-compute the expectation, you naturally obtain an expectation of

$\frac{1}{6} \cdot 6 + \frac{1}{2} \cdot 2 + \frac{1}{3}(-6) = 0$ coupons, so the expectation in the overall game is 0, indicating that it is fair (at least in theory) to the player and the house.

We now state the general procedure for computing mathematical expectation. Let

$$E_1, E_2, E_3, \ldots, E_n$$

be pairwise disjoint events (no pair can occur simultaneously) with

$$p_1, p_2, p_3, \ldots, p_n$$

their respective probabilities and

$$r_1, r_2, r_3, \ldots, r_n$$

their respective payoffs.

Then the "expected payoff" or *mathematical expectation* X in an experiment in which one of these n events must occur is defined by

$$X = p_1 r_1 + p_2 r_2 + p_3 r_3 + \cdots + p_n r_n.$$

Based on the first two examples and the idea behind this formula, we make the following observations about the expectation in a probabilistic experiment. The expectation need not be equal to any of the possible payoffs, but represents an appropriately weighted sum of the payoffs with the probabilities providing the weights. The sign of a given payoff reflects whether it is a gain or a loss, and care should be taken to make sure that losses contribute negative terms in the computation of X. Finally, a game or experiment is defined to be *fair* if its overall expectation is 0. We caution the reader again that the events $E_1, E_2, \ldots, E_n$ must be pairwise disjoint (no pair of events can occur simultaneously) and must exhaust the set of possible outcomes in order for the above expectation formula to be correct.

We conclude this chapter with several somewhat more realistic and significant applications of the expectation idea, which will be a recurring theme throughout the book.

Expectation example 3:   You are thinking of attending a late January convention whose nonrefundable registration fee is $15 if paid before January 1 (preregistration) and $20 if paid later or at the convention. You subjectively estimate on December 28 (last mailing date to insure arrival before January 1) that your probability of being able to attend the convention is p. For which values of p should you preregister and for which should you wish to pay at the door?

Solution: X(preregister) = $(1)(-15)$ (you pay \$15 in any event).
X (wait) = $p(-20) + (1-p) \cdot 0$ (you pay \$20 only if you attend).
These two expectations will be equal when $-15 = -20p$ or $p = 3/4$. Thus, for $p > 3/4$ you should preregister; for $p < 3/4$ you should wait; for $p = 3/4$ take your pick.

Expectation example 4:   (The futility of roulette) Suppose a \$1 bet is made on red with a Las Vegas roulette wheel. Then

$$X(\text{red}) = \frac{18}{38}(\$1) + \frac{20}{38}(-\$1) = -\frac{2}{38}(\$1) = -5.26 \text{ cents.}$$

Similarly, if the bet is on "1st third", then

$$X(\text{1st third}) = \frac{12}{38}(\$2) + \frac{26}{38}(-\$1) = -\frac{2}{38}(\$1) = -5.26 \text{ cents.}$$

If the bet is on a single number, then

$$X(\text{single \#}) = \frac{1}{38}(\$35) + \frac{37}{38}(-\$1) = -\frac{2}{38}(\$1) = -5.26 \text{ cents.}$$

In fact (see Exercise 2.7) any bet or combination of bets can similarly be shown to have the same negative expectation—you will lose an average of 5.26 cents for every dollar wagered, regardless of how it is wagered.

Expectation example 5:   (Pascal's wager) We somewhat facetiously analyze Blaise Pascal's philosophical argument (1660) that it is probabilistically prudent to believe in God. Let p = the probability that God exists and let us make the reasonable assumption that, whatever its value, $p > 0$ (the existence of God is not an impossible event). Each individual must make the decision either to believe or not believe in the deity. Reasoning that belief in a nonexistent God has some negative payoff (say $-z$) due to wasted time, energy, loyalty, etc. (in Pascal's translated words, the believer will "not enjoy noxious pleasures, glory, and good living"), but that nonbelief in the case where God exists has an infinite negative payoff (eternal damnation), we obtain the following table of payoffs, where x and y are some positive but *finite* payoffs:

	existence	nonexistence
belief	x	$-z$
nonbelief	$-\infty$	y

The expectations in cases of belief and nonbelief are

$$X(\text{belief}) = p \cdot x + (1 - p)(-z);$$

$$X(\text{nonbelief}) = p(-\infty) + (1 - p)y = -\infty.$$

We conclude that, regardless of how small the (assumed) positive probability of God's existence may be, the expectation is higher (perhaps less negative) if we choose to believe.[†] The reader may take this example as seriously as religion, faith in numbers, and sense of humor allow.

Exercises

2.1   Compute the probability of getting at least 1 six in 3 rolls of an honest die. Hint: first compute the probability of no six in 3 rolls. *Then expose the flaw in the following reasoning*:
The probability of a six in any one roll is $1/6$.
Since we have 3 rolls, $p(\text{at least 1 six}) = 3 \cdot \frac{1}{6} = \frac{1}{2}$, an even chance.

2.2   Solve Chevalier de Méré's problem of determining the probability of obtaining one or more double sixes in 24 rolls of a pair of dice. Hint: $(35/36)^{24} = .5086$. Conclude that de Méré was wise and sensitive to doubt the wisdom of an even-odds bet on this double six result.

2.3   Recall the "problem of points" treated early in the chapter, and explain how the stakes should be split in a 5 point game if Walter has 3 points and Jimmy 2 points when play is stopped.

2.4   Calculate the number of distinct, equally likely elementary events when 3 dice are rolled. Then compute $p(A), p(B),$ and $p(C)$ where $A =$ total of 3 or 4; $B =$ total of 5; $C =$ total of 6 or more. Finally, compute the odds for and odds against each of the three events A, B, and C.

2.5   If you are dealt $3, 4, 5, 7$ as the first four cards of a poker hand, what is

---

[†]An examination of Pascal's *Pensées* shows that he supposes the entry for nonbelief/existence to be finite while requiring the belief/existence entry to be positively infinite (eternal bliss). The author slightly prefers the version given here.

the probability of getting a 6 as the fifth card? If the fifth card is *not* a 6, compute the probability of replacing it (on a 1 card draw without replacement) with a new fifth card which is a 6. Obtain from this the general probability of drawing one card to fill an *inside straight*.

2.6   Compute the probability of 2 or more dice showing the same number of spots when 4 dice are rolled. Hint: first compute the probability of all 4 dice being different by looking at 4 successive independent trials.

2.7   Consider a Las Vegas roulette wheel with a bet of $5 on black and a bet of $2 on the specific group of 4 pictured in Figure 1 (13, 14, 16, 17). What is the bettor's expectation on this combined bet? Conclude, as suggested in the text, that the bettor's expected loss is 5.26 cents for every dollar bet.

2.8   A *European* roulette wheel has 37 sectors including a zero but no double zero. Furthermore if the zero comes up, any wager on *even money bets* (red, odd, 2nd 18, etc.) remains for *one more spin*. If the desired even money event occurs on this next spin, the original wager is returned (no gain or loss). Otherwise house odds are as listed in Table 2.
a)   Compute the expectation of a $1 bet on red with such a wheel.
b)   Compute the expectation of other types of $1 bets.
c)   Conclude that European roulette has a measure of "skill" not present in Las Vegas roulette.

2.9   A coin is flipped and then a single die is rolled. Payoffs in this successive game are as indicated below. What is the expectation of this game, and how much should be paid or charged to a player to make it a fair game?

	1 or 2	3, 4, or 5	6
H	12	− 10	0
T	− 6	9	− 6

2.10   A mythical slot machine has three wheels, each containing ten symbols. On each wheel there is 1 JACKPOT symbol and 9 other non-paying symbols. You put 1 silver dollar in the slot and the payoffs are as follows:
3 JACKPOT symbols—$487 in silver is returned (*including* your $1).
2 JACKPOT symbols—$10 in silver is returned.
1 JACKPOT symbol—$1 in silver is returned (you get your wager back).
Define what it would mean to say that this slot machine is *fair* and then show that it is indeed a fair one-armed bandit!

2.11   In backgammon a player's *count* on a given roll of two dice is determined as follows: If doubles are rolled the count is twice the total on the two dice (thus double 5 would give a count of 20). Otherwise the count is simply the total on the two dice. Compute the "expected" count on a backgammon roll. Your answer should be $8\frac{1}{6}$, but it is the method that is of interest here.

2.12   (St. Petersburg paradox) You flip a fair coin as many times as you can until you obtain a tail at which time you stop. The payoff to you is based upon the number of initial "heads" you get before you are stopped by a tail according to the table below. What is your expectation for this game? Discuss your answer; ask what it means and whether it is realistic.

# of "heads"	0	1	2	3	...	n	...
payoff (in dollars)	1	2	4	8	...	2^n	...

CHAPTER 3

Backgammon and Other Dice Diversions

Backgammon oversimplified

There are many good books on the rules and strategy of backgammon (see, for instance, Jacoby and Crawford's *The Backgammon Book*), and we do not attempt to duplicate these here. We assume the reader is either already familiar with the basic rules or else will leave us at this point to learn the game from a relative or friend. The rules and mechanics of backgammon are straightforward, but there is seemingly endless variety in the play of this delightful combination of skill and chance. We are primarily interested in those aspects of everyday back-gammon play which can be analyzed by means of the elementary probability and expectation techniques developed in the last chapter. Specifically, we shall consider probabilities of hitting *blots* (isolated opponents) in various situations, probabilities of entering from the bar and of *bearing off* (removing all men from the board to complete the game), and the beautiful mathematics of the doubling cube. Further-more, we shall pursue the application of some of this backgammon arithmetic in improving certain aspects of overall play. Figure 2 pro-vides a representation of the backgammon board and establishes some notation. We shall always assume that white is sitting in the "south" position and moving his men counterclockwise. The setup shown indi-cates how men are positioned at the start of a game.

Before considering specific situations, we present a simplified break-down of backgammon into its opening, offensive, defensive, and closing stages to convince the reader that there is much more to the game than quickly working one's way around the board. We leave details of the crucial doubling cube to a later section. The opening moves by each

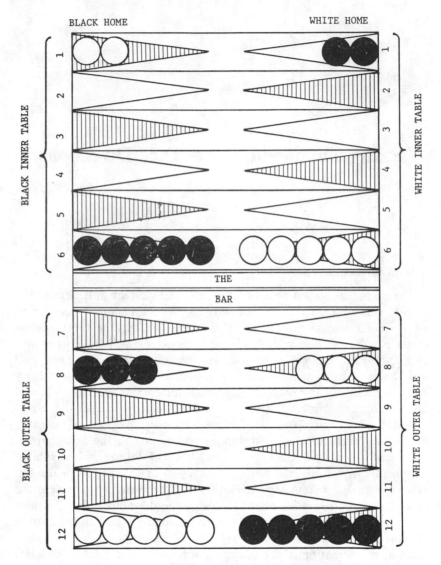

Figure 2   Schematic backgammon board and its reference points.

player have been given extensive treatment based primarily upon common sense and experience. Indeed it is virtually impossible to analyze mathematically the choices for given opening rolls, and we urge the reader to adopt one of the standard sets of recommendations for opening moves and responses (even here there is disagreement and some controversy). Table 4 lists the author's favorite opening moves, though

TABLE 4

Some Reasonable Opening Moves

DOUBLES—Assumes opponent went first and move is not blocked	
Roll	Recommended move
6-6	Capture both 7 points
5-5	Move two men from opponent's 12 point to your 3 point
4-4	Capture your 9 point and opponent's 5 point
3-3	Capture your 7 (alias *bar*) point
2-2	Capture opponent's 5 point
1-1	Capture your 7 (bar) point and your 5 point

NONDOUBLES—Assumes move is not blocked	
Roll	Recommended move
6-5 or 6-4	Move one back man (opponent's 1 point) the full count
6-3 or 6-2	Move back man to opponent's 7 point and advance man from 12 point
6-1	Capture your 7 (bar) point
5-4 or 5-2	Move two men from opponent's 12 point
5-3	Capture your 3 point
5-1	Advance one back man to opponent's 7 point
4-3	Move two men from opponent's 12 point
4-2	Capture your 4 point
4-1	Advance a man to your 9 point and a man to your 5 point
3-2	Move two men from opponent's 12 point
3-1	Capture your 5 point
2-1	Advance a man to your 11 point and a man to your 5 point

these vary depending upon mood and opponent's positioning (assuming opponent went first). Already at this early stage the player may feel the conflict between cautious, uninspired column-preserving moves and adventurous, blot-creating and point-building moves.

As the game develops each player concentrates, in varying degrees, upon

(a) moving his men safely and rapidly around to his inner table (the *running* game);
(b) blocking up his table with adjacent *points* (paired men) to prevent the opponent from escaping (the *blocking* game);
(c) holding two or preferably more men in the opponent's inner board in the hope of hitting the opponent at the right time (the *back* game).

Strategy (a) is the most straightforward and, played to the limit, often turns the game into one of pure dice-throwing and prayer. Successful employment of strategy (b) requires forethought, some chance-taking, awareness of the opponent's position, sound use of probabilities, and friendly dice. The back game strategy (c) tends to be a last (but most exciting and by no means hopeless) resort, to be employed when the opponent is well ahead (usually through more successful use of strategies (a) and (b)). A good player must be adept at knowing how and when to move among these three facets of the game, and this requires understanding of the odds on points being made, blots being hit, and men coming off the bar in various situations.

Finally, most good games of backgammon not terminated by the doubling cube require procedures for bearing off efficiently, and this too requires the ability to think and to count cleverly and quickly. In the following three sections we shall describe some of the elementary mathematics which justifies certain moves of skilled players (most of whom have somehow bypassed the mathematics with "game sense"). It is important to point out that the situations we shall consider will often be oversimplified and even unlikely. However, what we stress is the reasoning, because it, rather than the particular cases studied, will carry over to other situations. A systematic mathematical approach to the overall game is scarcely possible. As in chess, the number of choices and their ramifications at almost any stage is too large for human or machine analysis.

Rolling spots and hitting blots

The section of Chapter 2 on dice probabilities will be useful here, but it must be augmented slightly to take into account the special nature of rolling doubles in backgammon. Thus, if you are interested in moving a man to a position eight points away, recall that there are 5 ways for the two dice to total eight. Remembering also that double 2 will allow for a count of eight, we see that the probability of being able to move your man the desired eight points is 6/36 or 1/6 (assuming that the intermediate points are not blocked). Similarly, if you wish to move a specific man six points away, note that there are

5	ways to total six on two dice
1	way through double 2
11	ways involving a six on at least one of the dice.

Checking that these events are disjoint, we see that

$$p(\text{moving one man six points}) = \frac{17}{36},$$

almost an even chance. It should be clear that this kind of reasoning is vital in backgammon when one is faced with a choice of whether and where to leave a blot, and whether to play for hitting an opponent on the next turn.

Since hitting blots is so important, we include a table (again assuming no opposing points intervene) of probabilities for hitting a blot with a single man at varying distances. The construction of the full table proceeds just as in our computation above of the eight points away ($p = 1/6$) and six points away ($p = 17/36$) cases.

TABLE 5

Probabilities of Hitting a Single Blot with a Single Man

# of points away	# of ways to hit	Probability of hitting
1	11	11/36
2	12	12/36
3	14	14/36
4	15	15/36
5	15	15/36
6	17	17/36
7	6	6/36
8	6	6/36
9	5	5/36
10	3	3/36
11	2	2/36
12	3	3/36
15, 16, 18, 20, or 24	1 for each	1/36 for each

One conclusion from this table is fairly evident and is put to use by all competent backgammon players:

If you must leave a blot which you do not want hit, leave it at least 7 points from the threatening man if possible, in which case the farther the better (with 11 vs. 12 as the lone exception). If you must leave the blot within 6 points, move it as close as possible to the threat man.

We hasten to point out that this rule, like most, has some exceptions. Thus it is good practice if you must leave a blot to leave it where, should it escape being hit, it will do you the most good (potentially) on your next turn. It is also possible to use Table 5 to compute probabilities of hitting one of two blots and probabilities of a blot being hit by one of two men, etc. For instance, if an opponent leaves blots at distances two and seven from your man, there are 12 ways to hit the "two" blot and 6 ways to hit the "seven" blot. Since 2 of these 18 ways are counted twice

(2-5 and 5-2), there are 16 distinct ways to hit one or more of these blots. Note that we must not add the original probabilities (12/36 and 6/36) since the events are not disjoint, so we remove the overlap first (from either event) and then add to obtain an overall probability of 16/36 of making a hit. See Exercise 3.1 for more of this reasoning, but be prepared during a game to count from scratch rather than relying on excessive memorization of tables.

Entering and bearing off

An important part of the decision of whether and how exposed to leave a blot relates to the difficulty of entering (should disaster strike) from the bar. Thus the hitting of your blot in the very early game (first few moves) may hurt you in terms of overall count, but should not leave you unable to roll what is needed to enter quickly. As your opponent fills more points in his inner table this changes dramatically, as should your willingness to leave blots. The reader should have little trouble following the reasoning behind Table 6. For example, if 4 points are filled, then each die provides a probability of 4/6 of *not* entering. In independent successive rolls (probability rule 5),

$$p(\text{not entering}) = \frac{4}{6} \cdot \frac{4}{6} = \frac{16}{36} \quad \text{and thus} \quad p(\text{entering}) = 1 - \frac{16}{36} = \frac{20}{36}.$$

Again the reader can imagine and will encounter other situations such as trying to enter two men from the bar or trying to hit a blot in a partially filled inner table of the opponent. We leave these questions for the exercises.

TABLE 6

Entering from the Bar on a Given Roll

# of opponent's points in inner table	Probability of entering on next roll
0	1
1	35/36
2	32/36 = 8/9
3	27/36 = 3/4
4	20/36 = 5/9
5	11/36
6	0

The philosophy behind bearing off in a simple race to finish (no opponents on the bar, etc.) is, with some exceptions, straightforward.

Men should be brought into your inner table as economically as possible, a procedure which frequently loads up your six point heavily. Once all men are in the inner table, take as many men off as possible on each roll. The fun begins when relatively few men are left and a roll is made allowing men to be moved closer but not taken off. We illustrate with two examples, assuming in both that you want to maximize your chance of *winning on the following roll*.

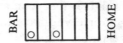

Backgammon example 1.   Your inner table is as indicated above, and you roll 3 and 2. Do you move to the 4 and 1 points or to the 3 and 2 points? If you move to 4 and 1 points only 7 rolls will prevent a win on the next turn (3-1, 1-3, 3-2, 2-3, 2-1, 1-2, 1-1). If you move to 3 and 2 points there are 11 rolls which will hurt (any roll including a "1"). Hence moving to the 4 and 1 points is the clearly superior play.

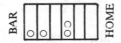

Backgammon example 2.   With your inner table as indicated above you roll 2 and 1. What do you do? The best play for finishing on the next turn is to move your 6 and 5 point men to the 4 point. Then you will finish with 4-4, 5-5, or 6-6. Any other move will require at least double 5 to finish. Note this is an exception to the principle of always removing a man if possible.

Questions involving bearing off against a back game or when an opponent is on the bar become more involved, but as always there is no substitute for careful counting when feasible and inspired thinking when possible (see Exercise 3.4).

The doubling cube

The incorporation of the doubling cube (in conjunction with a small stake) into a game of backgammon converts it from what might be a pleasant pastime to a dynamic, action-filled and fast-moving contest beautifully suited to gambling and strategic analysis. It is no coincidence that analysis of the doubling cube also involves some of the most elegant and applicable mathematics we will encounter. To summarize briefly the mechanics of the cube, each player has the initial option to double the stake of the game before one of his rolls. When a player doubles the stake, the cube is offered to the opposing player, who may either concede, paying out the stake existing before the double; or accept, controlling the cube for a possible subsequent double should his position call for this at some later point. Successive doubles alternate, with only the player controlling the cube having the option to double next.

It is important to remember that control of the doubling cube provides a dual advantage. First, you have the option to double if you wish. In addition, the opponent is unable to double and thus force you to concede when you are well behind. This advantage should not be used as a justification for accepting doubles simply to control the cube, though it will be used to suggest that you should not double the cube away with only a slight advantage.

We now consider when you should accept a double and when you should double. To deal with this mathematically we assume the current stake of the game is s units (the value of s will turn out to be immaterial to the final conclusions). We further assume that at the time a decision (either for accepting or doubling) is to be made

$$p = \text{your estimated probability of winning the game,}$$

$$1 - p = \text{your estimated probability of losing the game.}$$

Clearly, exact determination of p is next to impossible except in very simple end game situations, though there are useful rules of thumb for estimating p (see Jacoby and Crawford, pp. 106–108). We suppose, however, that an experienced player can estimate his probability of winning at any stage with reasonable accuracy.

Suppose you are doubled under the conditions assumed above. Then

$X(\text{refuse}) = -s$	(you lose s units)
$X(\text{accept}) = p(2s) + (1 - p)(-2s) = s(4p - 2)$	(the payoff is up to $2s$ units).

You should accept when $X(\text{accept}) > X(\text{refuse})$ or when $s(4p - 2) > -s$. Dividing both sides of this inequality by the positive common factor s, you should accept when $4p - 2 > -1$. Thus, *accept a double when $p > 1/4$.* Similar reasoning indicates that you should *refuse when $p < 1/4$* and follow your whim when $p = 1/4$. A more delicate analysis taking into account the fact that accepting gives you control of the cube might show, in certain cases, that doubles could be accepted with mathematical justification for values of p somewhat less than $1/4$. On the other hand, if possibilities of being "gammoned" or "backgammoned" exist, you might want p to be somewhat greater than $1/4$ before accepting.

Consider the opposite question of when a double should be given to your opponent. Under what circumstances do you *want* him to accept your double? Working again with expectations, with p and s as above,

$$X(\text{opponent accepts}) = p(2s) + (1-p)(-2s) = s(4p-2);$$

$$X(\text{opponent refuses}) = s.$$

Equating these expectations and solving for p, we obtain $p = 3/4$ as the probability at which you do not care whether or not your double is accepted. If $p < 3/4$ you prefer a concession, and if $p > 3/4$ you hope your foolish opponent accepts the double. Now consider a third expectation, namely

$$X(\text{no double}) = p(s) + (1-p)(-s) = s(2p-1).$$

Since $X(\text{no double}) = X(\text{opponent accepts})$ precisely when $p = 1/2$ (check this), we make the following preliminary conclusions: Whenever $p > 1/2$, doubling *seems* a good idea; and if $p > 3/4$, you should hope the double is accepted. When $p < 1/2$ doubling is a bad proposition $[X(\text{no double}) > X(\text{opponent accepts})]$ unless your opponent is timid enough to refuse such doubles $[X(\text{opponent refuses}) > X(\text{no double})]$. After all this hard work, we remind ourselves that doubling gives up control of the cube, a fact not easily fitted into the mathematical analysis. But intuitively, it suggests that doubles should not be made (except in end game situations) unless p is safely greater than $1/2$ (say $p \geqslant 2/3$). If neither player has doubled, your double is not giving the opponent much more control than he already had, so *initial* doubles might be made with p shading even closer to $1/2$ from above.[†] Finally, we caution that doubling when you have a good chance at a gammon should be avoided even if p is well over $1/2$ (paradoxically, in these cases the larger the p value, the less you should be inclined to double).

[†] Keeler and Spencer, in their paper "Optimal Doubling in Backgammon" [*Operations Research* 23, 1063–1071 (1975)], give a detailed mathematical analysis of doubling. Ignoring gammons and backgammons, they indicate that optimal strategy calls for a double when $p > 4/5$ in the early game, with p dropping towards $1/2$ as the end game is approached. They also suggest accepting early doubles whenever $p > 1/5$.

We conclude our backgammon discussion with several more examples illustrating the ideas we have developed.

Opponent inner table

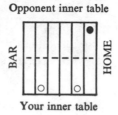

Your inner table

Backgammon example 3. It is your roll with your (white) inner table and your opponent's as shown above. Should you double? You have 17 losing rolls—11 with "1" on either die, 3-2, 2-3, 4-2, 2-4, and 4-3, 3-4. Hence your probability p of winning is $p = 19/36$. You are not worried about losing control of the cube (if you fail on this turn you lose anyway!). Hence you should double since $p > 1/2$, and opponent should accept since $1 - p > 1/4$ (recall that $1 - p$ is opponent's probability of winning).

Opponent inner table

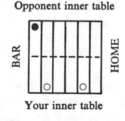

Your inner table

Backgammon example 4. Again it is your turn (white) with the board as shown above and you in control of the cube. Now should you double? Three disjoint events can take place:

A = you win on your first roll

B = opponent wins on his first roll (which requires *not A* to happen first)

C = you win on your second roll (which requires *not A* and then *not B* to happen first).

(We ignore the extremely unlikely event that you will roll 2-1, opponent will not finish, and then you roll 2-1 again—you would double after opponent failed and be in situation *C*.)

$p(A) = \dfrac{19}{36}$ as in example 3.

$p(B) = \dfrac{17}{36} \cdot \dfrac{3}{4}$ (only 9 rolls will prevent opponent from finishing).

$p(C) = \dfrac{17}{36} \cdot \dfrac{1}{4}$.

As the size of the stake is immaterial to our analysis, we assume a stake of 1 unit. Then,

$$X(\text{no double}) = \frac{19}{36}(1) + \frac{17}{36} \cdot \frac{3}{4}(-1) + \frac{17}{36} \cdot \frac{1}{4}(1) = \frac{21}{72},$$

$$X(\text{double, accept redouble}) = \frac{19}{36}(2) + \frac{17}{36} \cdot \frac{3}{4}(-4) + \frac{17}{36} \cdot \frac{1}{4}(4) = \frac{8}{72}.$$

[Note that the payoffs of 4 result from the fact that opponent will redouble].

$$X(\text{double, refuse redouble}) = \frac{19}{36}(2) + \frac{17}{36}(-2) = \frac{8}{72}.$$

Comparing these three expectations we see that the best course is to hold off on the double because of the threat of a redouble, illustrating the value of controlling the cube. Note the paradoxical fact that you should double in example 3 with the opponent on his one point, but refrain in example 4 despite the fact that the opponent is farther back (on his six point).

Opponent inner table

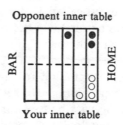

Your inner table

Backgammon example 5.   You have the cube and the dice as white with the board shown above. What should you do? Your probability of winning the game is at least 31/36 since you will surely be done in 2 turns and opponent has only 5 rolls (any double except 1-1) to prevent you from getting a second turn. (Since you may roll doubles yourself, things are even better than this lower estimate. A complete calculation shows that $p(\text{you win}) = 1 - \frac{31}{36} \cdot \frac{5}{36}$.) You should certainly double, in which case opponent (with his probability of winning less than 1/4) should resign. Note that a failure to double here gives the opponent a chance to win a game that your double should guarantee for yourself.

As is often the case with a mathematically based analysis of a rich and dynamic game situation, crucial psychological considerations have been unjustifiably ignored. In the case of the doubling cube, an opponent should be tested with a variety of doubles, including a few mathematically unsound ones. Indeed your opponent's estimate of $1 - p$ (his probability of winning) may tend to be unduly conservative, in which case a forceful doubling policy is called for. Similar considerations involving acceptance of doubles, rationally employed, can add a psychological dimension to your game which can, in certain situations, be far more effective and profitable than unwavering mathematical optimization.

Craps

One of the mainstays of American casino gambling craps is a game with a simple set of basic rules and a dizzying variety of options involving bets and their resultant odds. Craps is also a common street (or locker room, back alley, etc.) gambling game in which players bet against each other rather than against a monolithic "house". We shall first describe and analyze *street craps* and then consider the delicate alterations which the gambling houses make in offering *casino craps* to their customers.

In street craps the shooter rolls two dice. If a total of 7 or 11 comes up on a beginning roll (an event called a *natural*), the shooter and those betting with him (called *pass* bettors) win the amount bet. If a 2, 3, or 12 total (called a *craps*) shows on the beginning roll, the pass bettors lose and the competing *don't pass* bettors win. Any of the remaining six totals (4, 5, 6, 8, 9, 10) on the beginning roll becomes the pass bettors' *point*. In this case, the shooter continues to roll the dice and if the point comes up before a seven, the pass bettors win. Otherwise (i.e., a seven comes up before the point), the don't pass bettors win. In this "friendly" craps game any bet made by a player must be "covered" by a player betting on the opposite result. Rolling and betting start again after each pass or don't pass, with the shooter keeping the dice as long as he makes his point, rolls naturals, or rolls craps, and passing the dice upon losing his point.

To increase the action there are all sorts of side bets made at appropriate odds. Before considering these options we analyze the expectation on a simple pass or don't pass bet. The probabilities of naturals and craps can be computed directly by counting as done previously. Probabilities for making and losing points involve a small extra step; we illustrate as follows: Suppose *five* is the point to be made (it came up on a starting roll). There are 4 ways to total five (pass) and 6 ways to total seven (don't pass). Thus with five as the point, $p(\text{pass}) = 4/10$, the odds being 6:4 or 3:2 against. The probability of rolling 5 $(4/36)$ and *then* passing is $(4/36) \cdot (4/10) \approx .0444$. Similarly $p(\text{point of 5}$ then don't pass$) = (4/36) \cdot (6/10) \approx .0667$. By using this reasoning we can compute in Table 7 the probabilities of the various disjoint events making up the dice rolling experiment known as craps.

TABLE 7

Events and Probabilities in a Play of Street Craps

Initial roll	p(Initial roll)	p(Initial roll then pass)	p(Initial roll then don't pass)
natural (7 or 11)	$\frac{8}{36}$	$\frac{8}{36} \approx .2222$	$0 = .0000$
craps (2, 3, or 12)	$\frac{4}{36}$	$0 = .0000$	$\frac{4}{36} \approx .1111$
4	$\frac{3}{36}$	$\frac{3}{36} \cdot \frac{3}{9} \approx .0278$	$\frac{3}{36} \cdot \frac{6}{9} \approx .0556$
5	$\frac{4}{36}$	$\frac{4}{36} \cdot \frac{4}{10} \approx .0444$	$\frac{4}{36} \cdot \frac{6}{10} \approx .0667$
6	$\frac{5}{36}$	$\frac{5}{36} \cdot \frac{5}{11} \approx .0631$	$\frac{5}{36} \cdot \frac{6}{11} \approx .0757$
8	$\frac{5}{36}$	$\frac{5}{36} \cdot \frac{5}{11} \approx .0631$	$\frac{5}{36} \cdot \frac{6}{11} \approx .0757$
9	$\frac{4}{36}$	$\frac{4}{36} \cdot \frac{4}{10} \approx .0444$	$\frac{4}{36} \cdot \frac{6}{10} \approx .0667$
10	$\frac{3}{36}$	$\frac{3}{36} \cdot \frac{3}{9} \approx .0278$	$\frac{3}{36} \cdot \frac{6}{9} \approx .0556$
Totals	1	$= .4928$	$= .5071$

The totals in the last two columns essentially add to 1 (the discrepancy results from roundoff errors) and indicate that the don't pass bet has a slight edge. Indeed, one will see "street wise" players betting don't pass consistently except when they are shooting, at which time matters of personal involvement and force of will override these delicate mathematical considerations. To see precisely what to expect, we assume a $1 bet and use the expectation concept.

$$X(\text{pass}) \approx .4928(1) + .5071(-1) = -.0143 \quad [-1.4 \text{ cents}]$$

$$X(\text{don't pass}) \approx .4928(-1) + .5071(1) = .0143 \quad [+1.4 \text{ cents}]$$

The concept of a positive expectation for the bettor is of course intolerable for a casino, and the means of converting street craps into a viable casino game is interesting. The rules for pass betting are identical. For don't pass betting an initial craps roll of double six (or in some places double one) still loses for the pass bettor, but the don't pass bettor neither wins nor loses. (A few casinos use "total of 3" as the no result event—see Exercise 3.9). The probability of an initial roll of double six is $1/36$. Under street rules this outcome contributes a term of $(1/36) \cdot 1$ to the expectation of a don't pass bettor. But under casino rules it contributes a term of $(1/36) \cdot 0 = 0$ to his expectation. Therefore, the positive expectation of .0143 for the don't pass bettor on the street is decreased by $(1/36) \approx .0278$ in the casino. Consequently, for casino craps we have

$$X(\text{don't pass}) \approx .0143 - .0278 = -.0135 \quad [-1.4 \text{ cents}].$$

This strategem by the casinos enables them to field both pass and don't pass bets with similar mathematical confidence. Indeed the house expects to win about 1.4 cents on every dollar wagered on either bet. The fact that this *house edge* of 1.4 percent is less than the edge in roulette by a factor of 4 may partially explain the relative popularity of craps among the "smart" American casino gamblers.

We will not discuss all the betting options available in craps, though Figure 3 should suggest the richness of choice. We consider a few mathematical aspects of these options in the exercises and treat here the option of most interest to the expectation-conscious bettor, that of *free odds*. If the initial roll is neither a natural nor a craps, so that a point is established, a bettor may "back up" his bet with up to an equal additional amount at true mathematical odds. Thus a pass bettor who bets $10 and sees a point of 4 established is in a disadvantageous position, having only 3 chances out of 9 to win at this point. He may, however, stake an additional $10 on 4 to appear before 7 at the true $2:1$ odds against, thereby improving his *per dollar* expectation in his current unfavorable position. If he takes these free odds and the 4 point is made, he will win $30 overall, while the rolling of a now dreaded 7 will

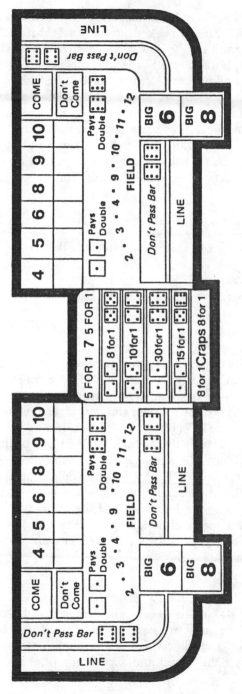

Figure 3   A casino craps table layout. A LINE bet is a bet to pass. Note that the Don't Pass bet loses on an initial double six. Come and Don't Come bets are analogous to Pass and Don't Pass bets except that they can be initiated any time rather than only when a new shooter is coming in. The BIG 6 and BIG 8 bets are even money that the indicated number will come up before a seven. See Exercises 3.7 and 3.8 for explanation and analysis of some of the other bets.

lose him $20 overall rather than his initial $10. Similarly a don't pass bet of $6 and a point of 8 can be backed up with a $6 bet at 5:6 odds against that the point will not be made. The reader should check that $12 is lost if the point is made and $11 is won otherwise.

Let us analyze the optimal craps strategy of betting $10 on pass and taking maximal free odds whenever possible (actually a similar strategy with a $6 bet on don't pass would be slightly better—why?). Since the house will not pay out fractions of some minimal stake, such as $1, it is necessary to make bets in multiples of $10 to receive full payment on this free odds strategy. Using Table 7, the various payoffs on free odds, and the ubiquitous expectation formula, for this strategy we have on an initial $10 bet

$$X(\text{pass}) \approx \underbrace{.2222(10)}_{\text{natural}} + \underbrace{.1111(-10)}_{\text{craps}} + \underbrace{.0556(30)}_{\substack{\text{pass on}\\ \text{4 or 10}}} + \underbrace{.1111(-20)}_{\substack{\text{don't pass}\\ \text{on 4 or 10}}}$$

$$\underbrace{+ .0889(25)}_{\text{etc.}} + .1334(-20) + .1262(22) + .1514(-20)$$

$$\approx -.145 \qquad\qquad\qquad [-14.5 \text{ cents}].$$

At first glance it appears that this aggressive strategy barely affects the bettor's welfare (and in fact slightly reduces it). If we realize, however, that this expectation results from increased betting, we see it as a better percentage bet. Specifically, the average amount wagered under our free odds, pass bet strategy is $(1/3)(10) + (2/3)(20) \approx 16.67$ dollars (1/3 of the time the initial roll decides the bettor's fate and 2/3 of the time he backs up his point). Thus this strategy has an expectation *per dollar wagered* of $-.145/16.67 \approx -.0087$ [$-.9$ cents/dollar] and the house edge has been cut below 1 percent. Not only will much of the "smart" money in casino gambling be found around the craps tables, but the smartest of the smart money will be taking advantage of the free odds. Table 8 lists the various types of bets available in craps and the edge each provides to the house.

Chuck-a-Luck

The game of chuck-a-luck is nearly an ideal carnival fixture. The apparatus is simple, the rules are easy to understand, the odds *seem* attractive to the bettor, and the carnival makes money steadily. In this brief section we analyze the appeal and expectation of chuck-a-luck.

TABLE 8

Odds and House Edge on Craps Bets (Best to Worst)

Type of bet	Probability of win	House odds	House edge
Pass/Don't Pass (plus free odds)	—	—	.9%
Pass/Don't Pass (no free odds)	$\approx .493$	1 : 1	1.4%
Field	$\frac{16}{36} \approx .444$	—	5.6%
Big 6/Big 8	$\frac{5}{11} \approx .455$	1 : 1	9.1%
Double 3/Double 4	$\frac{1}{11} \approx .091$	9 : 1	9.1%
Double 2/Double 5	$\frac{1}{9} \approx .111$	7 : 1	11.1%
Craps	$\frac{4}{36} \approx .111$	7 : 1	11.1%
Double 1/Double 6	$\frac{1}{36} \approx .028$	29 : 1	16.7%
1-2/5-6	$\frac{1}{18} \approx .056$	14 : 1	16.7%
Seven	$\frac{1}{6} \approx .167$	4 : 1	16.7%

The apparatus consists of three dice (usually large ones in a wire cage). The betting surface consists simply of regions containing the numbers 1 through 6. The bettor places his stake in a region containing a particular number, and the three dice are rolled once. The bettor wins the amount of the stake for each appearance of the number bet upon. If the number does not appear, the stake is lost. For example, if $1 is placed on *five* and the spots 5-3-5 come up the bettor wins $2 (collecting $3 including the original stake). Similarly 4-4-1 would lose $1, 6-2-5 would win $1, and 5-5-5 would win $3.

This game seems attractive, even to some who think intuitively about probabilities, because one might reason as follows: There are three dice and six possible faces on each die, so any given number should have a 3/6 or an even chance of appearing, ensuring at least a fair game. Furthermore, the possibility of doubles and triples yields extra money to be won. This is a nice example of plausible but false probabilistic reasoning (recall Exercise 2.1 and the need for disjoint events in rule 4 for probabilities). As always, a careful analysis must be made of the situation.

There are $6^3 = 216$ equally likely elementary events in rolling three fair dice. Exactly one of these pays triple, so p(triple win) $= 1/216$. To count the number of elementary events leading to a double win, assume for specificity that a bet is placed on 4. The desired elementary events have the form 4-4-x, 4-x-4, or x-4-4 where x in each case denotes "not 4" and has 5 ways of occurring on a particular die. Thus there are $5 + 5 + 5 = 15$ ways to roll exactly two 4's. Similarly, single win elementary events have the form 4-x-y, x-4-y, or x-y-4 where both x and y represent "not 4." Each of these 3 single win methods can happen in $5^2 = 25$ ways (x and y each have 5 possibilities and vary independently). Subtracting $1 + 15 + 75$ from 216 to obtain 125, we are ready to compute an expectation—as usual on a $1 bet.

$$X(\text{chuck-a-luck}) = \frac{1}{216}(3) + \frac{15}{216}(2) + \frac{75}{216}(1) + \frac{125}{216}(-1)$$

$$= \frac{(108 - 125)}{216} = \frac{-17}{216} \approx -.0787 \; [-7.9 \text{ cents}].$$

The carnival has an expectation of 7.9 cents profit on every dollar bet, better than roulette by a factor of about 1.5. The appearance of the figure 108 ($216 \div 2$) in our calculation above helps to explain the intuitive appeal of chuck-a-luck to the bettor. In 216 rolls of the three dice the bettor will win, on the average, 108 dollars (on 91 of the rolls) while losing 125 dollars on the other 125 rolls. It is no mystery that carnivals cherish this aptly named game.

Exercises

(Exercises 1 to 6 refer to Backgammon)

3.1   a) You leave two blots which are 4 points and 6 points from one of your opponent's men (no other men in sight). What is the opponent's probability of being able to hit at least one of your men on the next roll?

b) Suppose conditions are as above except that you have also established a point 3 points away from your opponent's man. Now what is the probability of one or more of your blots being hit on the next roll?

3.2   a) Suppose you have two men on the bar and two points are held by the opponent in his inner table. Show that the probability of entering both of your men is 4/9 and that the probability of entering at least one of your men is 8/9. Use this to compute the probability of entering exactly one of your two bar men.

b) Construct a table analogous to Table 6 where you have two men on the bar and the table's columns are labeled:

Number of opponent's points in inner table,

p(Entering both men on next roll),   and

p(Entering at least one)

3.3   You have one man on the bar. Opponent controls his 2, 4, and 6 points and has a blot on his 5 point. It is your turn. What is the probability of being able to hit the blot on your next roll? What is the probability of your having to enter from the bar without hitting the blot?

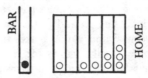

3.4   a) Your inner table and all your remaining men are shown above. It is your roll; opponent (black) has one man on the bar and at least 8 other men in his inner table. You roll 6-3. What is your best play?

b) The situation is identical to the one described above except that the opponent's man on the bar is his only remaining piece. Now what is your best play?

3.5   You have a single man remaining, 8 points from home, and it is your roll. Your opponent will surely go out if he gets another turn. Show that p(your winning) $< 1/2$, so that a double is not called for. Then explain why this does not contradict the result stated in Exercise 2.11— namely that the expected count on a backgammon roll is $8\frac{1}{6}$, exceeding the 8 points you need to win

3.6   Imagine a backgammon game with the doubling cube replaced by a "tripling cube" (with faces of 3, 9, 27, 81, 243, 729). Following the analysis given for the doubling cube, compute the probability of winning above which a triple should be accepted.

(Exercises 7, 8, and 9 refer to Craps)

3.7   (See Figure 3 and Table 8) A *field* bet is a bet that a total of 2, 3, 4, 9, 10, 11, or 12 will come up on the next roll. Double 1 and double 6 pay 2 to 1, all others paying even odds. Compute the bettor's expectation on a $1 field bet and thus obtain the percentage house edge on a field bet. Check your results with Table 8.

3.8   Using Figure 3, compute the house edge (in percent) for each bet whose rules and payoff make sense to you from the figure. A *hardway* 8 for 1 bet on double 2, for example, pays 8 *for* 1 (not 8 to 1) and wins when the indicated roll (double 2) comes up before a seven or a non-doubles four total. The bet stays out until a four or seven total comes up. The 10 for 1 hardway bets are similar. The longshot bets (30 for 1 and 15 for 1) are single roll bets that the indicated total will come up on that roll. You bet double 1 *by itself*, and not in conjunction with the double 6.

3.9   Consider a casino in which the "no result" craps roll for don't pass bettors is "total of 3." Compute the house edge (in percent) on a don't pass bet. Would you prefer to bet pass or don't pass in a craps game at such a casino?

3.10 In the standard chuck-a-luck game doubles pay $2:1$ and triples pay $3:1$. Imagine a rules change in which doubles pay $x:1$ and triples pay $y:1$. Derive an equation involving x and y (as unknowns) which describes the relationship between x and y that makes this modified chuck-a-luck a *fair game*. Give some specific values for x and y which satisfy this equation and propose a "best" solution in terms of ease of payoff, ease of memory, and mathematical appeal.

CHAPTER 4

Permutations, Combinations, and Applications

Careful counting: Is order important?

Much of the challenge in computing probabilities in situations with a finite number of elementary events consists of organizing information systematically, counting cases carefully, and applying the rules correctly. In this chapter we focus on the organization and counting process, applying it to some old and new game situations. A crucial concept in what follows will be the distinction between a *permutation* and a *combination* of objects (people, horses, cards, letters, or other entities). A permutation is a selection of objects in which *order* is taken into account. A combination is a selection in which order is unimportant. To motivate and clarify these concepts we ask and then answer five typical counting questions.

Question 1. In how many different ways can a bridge player arrange a particular (13 card) bridge hand?

Question 2. How many different finishes among the first 3 places can occur in an 8 horse race (excluding the possibility of ties)?

Question 3. How many foursomes can be formed from among 7 golfers?

Question 4. How many different 13 card bridge hands are there?

Question 5. How many different baseball batting orders are possible with a 25 player roster in which any player can bat in any position?

We answer these questions now, starting out slowly and picking up speed.

Solution 1. If we think of ordering the given hand from positions 1 through 13, there are 13 choices for the first position, then 12 choices for the second, then 11 for the third, and so on down to 2 choices for the twelfth position with the 1 remaining card going in the last position. Thus there are

$$13 \cdot 12 \cdot 11 \cdot \ \cdots \ \cdot 3 \cdot 2 \cdot 1 = 6,227,020,800$$

arrangements of a given bridge hand. Why do we multiply above rather than add?

Solution 2. There are 8 choices for the winner, then 7 for the place horse (2nd place), then 6 for the show horse (3rd place). Accordingly there are $8 \cdot 7 \cdot 6 = 336$ different orders of finish as far as the payoff windows are concerned (ignoring ties).

Solution 3. Consider a score card signed by the foursome. There are 7 possibilities for the first name on the card, six possibilities for the 2nd name, then 5 possibilities for the 3rd and 4 possibilities for the last name. This gives $7 \cdot 6 \cdot 5 \cdot 4 = 840$ possibilities, where *different orderings* of the same 4 players have been counted as *different foursomes*. But the wording of the question indicates that order is unimportant, so we realize that the result of 840 contains considerable duplication. In fact a given foursome can be ordered in $4 \cdot 3 \cdot 2 \cdot 1 = 24$ different ways, so our total of 840 has counted each foursome 24 times. We see then that there are $840/24 = 35$ possible foursomes that can be chosen.

Solution 4. Again order is unimportant here. We wish to know how many different hands of 13 cards can be chosen from a 52 card deck. If order were important there would be

$$\underbrace{52 \cdot 51 \cdot 50 \cdot 49 \cdot 48 \cdot \ \cdots \ \cdot 41 \cdot 40}_{\text{13 factors}}$$

possibilities. Since order is unimportant we must divide by $13 \cdot 12 \cdot 11 \cdot \ \cdots \ \cdot 3 \cdot 2 \cdot 1$. We thus obtain the smaller but still astronomical count of

$$\frac{52 \cdot 51 \cdot 50 \cdot \ \cdots \ \cdot 40}{13 \cdot 12 \cdot 11 \cdot \ \cdots \ \cdot 1} = 635,013,559,600$$

different bridge hands.

Solution 5. Here the wording of the question demands that order be taken into account, so we do not factor out duplications. The answer is

$$25 \cdot 24 \cdot 23 \cdot 22 \cdot 21 \cdot 20 \cdot 19 \cdot 18 \cdot 17 = 741,354,778,000$$

9 factors

a rather large number of batting orders.

Two very important characteristics of the above solutions should be noted. First, all of them involve *choosing without replacement* some subset from a set of objects. Thus, as each choice is made (position is filled) the number of choices for the next position decreases by one. This accounts for the repeated appearance of consecutive integers (decreasing by 1) and motivates the notation developed in the next section. If our questions had involved *choosing with replacement* such as asking for the number of different 3 spin sequences on a Las Vegas roulette wheel, the factors multiplied would have remained constant rather than decreasing by one each time.

A second aspect of our questions is that those asking for ordered subsets (permutations) have one general form of answer, while those where order is unimportant (combinations) have a related form which requires dividing out duplications. The decision as to which approach is called for requires a careful reading of the question and some realistic thought about the problem at hand. Specifically, questions 1, 2, and 5 involve permutations, while 3 and 4 call for combinations. We cannot overemphasize the necessity of thinking carefully about whether order is important in a given situation before plunging ahead.

Factorials and other notation

To deal with products of successive integers we introduce some standard and very convenient notation. We abbreviate the product appearing in solution 1 by 13! (read thirteen *factorial*) and, more generally, we define for any positive integer n,

$$n! = n(n-1)(n-2)(n-3)\ldots3\cdot2\cdot1.$$

Thus $3! = 6$, $5! = 120$, and $6! = 720$. As a very useful convention we also *define* $0! = 1$. In working with factorial notation it is most helpful to note that, for instance, $6! = 6 \cdot (5!)$, so that subsequent factorials can be computed easily from their predecessors. More generally,

$$(n + 1)! = (n + 1)(n!).$$

As can be seen by reviewing the solutions in the previous section, a single factorial is not usually sufficient to express the desired answer (this happened only for question 1). A careful look at the massive cancellation that can occur when factorials divide one another should show that products and quotients of factorials will serve admirably. Specifically, the permutation questions 1, 2 and 5 have answers

$$\frac{13!}{(13-13)!} = 13!, \quad \frac{8!}{(8-3)!} = \frac{8!}{5!}, \quad \text{and} \quad \frac{25!}{(25-9)!} = \frac{25!}{16!}.$$

The combination questions 3 and 4 have, with the division for duplications, respective answers

$$\frac{7!}{4!\,(7-4)!} = \frac{7!}{4!3!} \quad \text{and} \quad \frac{52!}{13!\,(52-13)!} = \frac{52!}{13!39!}.$$

(The factorially inexperienced reader should now check that these 5 quotients do indeed agree with the answers given in the previous section.)

We see from the above that all questions involving ways of choosing some number (say r) of objects from an overall number (say n) of objects can be answered by means of two types of formulas involving factorials, one formula for permutations and another for combinations. To highlight this and provide us with even more notational convenience we define

$$P_{n,r} = \frac{n!}{(n-r)!}$$

   [The number of *permutations* of n objects taken r at a time],

$$C_{n,r} = \frac{n!}{r!\,(n-r)!}$$

   [The number of *combinations* of n objects taken r at a time].

For instance, the number of combinations of 5 things taken 2 at a time is

$$C_{5,2} = \frac{5!}{2!3!} = \frac{5\cdot4\cdot3\cdot2\cdot1}{2\cdot1\cdot3\cdot2\cdot1} = 10.$$

The formula is easy to remember, and cancellation of the largest factorial in the denominator leads to simple computations.

We illustrate the use of this shortcut in writing and reasoning with two additional examples before looking at some more realistic applications in the next sections.

Question 6.   In how many different ways can 3 zeros come up in 10 spins of a roulette wheel?

Solution.   A typical configuration for the three zeros might be the 1st, 5th, and 7th spins of the wheel. This is a combination question involving 10 objects taken 3 at a time. Indeed, if the 10 spins are represented by 10 slots, we are interested in the number of sets of 3 slots which can be filled by zeros. By the wording of the question we do not care what nonzero numbers appear in the remaining 7 slots. Order is unimportant here since, for instance, switching the zeros in the 1st and 5th spins does not result in a different event. Hence the desired number of *combinations* is

$$C_{10,3} = \frac{10!}{3!7!} = \frac{10 \cdot 9 \cdot 8}{3 \cdot 2 \cdot 1} = 120.$$

Question 7.   How many seating arrangements are possible at a poker game for 8 where the host and hostess must sit at the ends of a narrow, long, rectangular 12 seated table (1 chair at each end)? The vacant chairs must remain at the table.

Solution.   Assuming the host and hostess have seated themselves, there are 10 seats remaining to be filled by the 6 guests. Clearly order is important, so the guests can be seated in $P_{10,6} = 10!/4!$ different ways. The host and hostess have 2 choices, so there are $2 \cdot 10!/4! = 302,400$ seating arrangements.

Probabilities in poker

Poker, like backgammon, stands on the borderline between games of chance (roulette, craps, chuck-a-luck, etc.) and games of pure skill (checkers, chess, go, etc.). As with backgammon we do not propose a thorough analysis of poker, referring the interested reader to books which attempt to treat the theory and practice in detail. Our discussion here will deal with special cases involving three aspects of the game. We first illustrate the utility of combinations in computing the probabilities of some of the well-known 5 card poker hands. Next we analyze some more realistic questions about obtaining certain hands in the midst of a deal or on a draw. Finally, we apply some oversimplified assumptions to questions of betting—when to drop, when to call, and when to raise. As usual, a mastery of the specific results we consider may not lead directly to more profitable poker playing, but should give increased insight into the conscious or unconscious thought process employed by those "lucky" consistent winners possessed of shrewdness, sensitivity and good instincts.

Ignoring the order in which the cards are dealt (unimportant in draw poker though crucial in stud where some of the cards are dealt face-up), we have

$$C_{52,5} = \frac{52!}{5!47!} = 2{,}598{,}960$$

equally likely 5 card poker hands. The probability of getting any one hand is then $1/C_{52,5}$ (about one in $2\frac{1}{2}$ million). Early in Chapter 2 we computed, by multiplying probabilities, the probability of being dealt 5 Spades in 5 cards. To illustrate the use of combinations we do this another way. There are

$$C_{13,5} = \frac{13!}{5!8!} = 1287$$

ways of getting 5 Spades, so

$$p(5 \text{ Spades}) = \frac{C_{13,5}}{C_{52,5}} = \frac{13!47!}{8!52!} = \frac{13 \cdot 12 \cdot 11 \cdot 10 \cdot 9}{52 \cdot 51 \cdot 50 \cdot 49 \cdot 48} = .000495,$$

agreeing with our previous answer. If we realize that of the 1287 Spade hands 10 are *straight flushes* (like 8, 9, 10, Jack, Queen of Spades), leaving 1277 mere flushes, and that the same applies to each of the other suits, then we can find

$$p(\text{flush}) = \frac{4 \cdot 1277}{C_{52,5}} = .0019654 \qquad \left[\approx \frac{1}{509}\right]$$

$$p(\text{straight flush}) = \frac{40}{C_{52,5}} = .0000153 \qquad \left[\approx \frac{1}{65{,}359}\right].$$

We apply these methods to other highly regarded hands; an Aces over Kings full house (AAAKK) can occur in $C_{4,3} \cdot C_{4,2}$ ways and there are $13 \cdot 12$ possible types of full houses (why?). Hence

$$p(\text{full house}) = \frac{13 \cdot 12 \cdot C_{4,3} \cdot C_{4,2}}{C_{52,5}} = \frac{13 \cdot 12 \cdot 4 \cdot 6}{C_{52,5}} = .00144 \qquad \left[\approx \frac{1}{694}\right].$$

By similar reasoning, which the reader is urged to reconstruct,

$$p(4 \text{ of a kind}) = \frac{13 \cdot 48 \cdot C_{4,4}}{C_{52,5}} = \frac{13 \cdot 48 \cdot 1}{C_{52,5}} = .00024 \quad \left[\approx \frac{1}{4167} \right].$$

Also

$$p(\text{straight}) = \frac{10 \cdot 4^5 - 40}{C_{52,5}} = \frac{10,200}{C_{52,5}} = .00392 \quad \left[\approx \frac{1}{255} \right].$$

(In this last calculation, there are 10 types of straights and 4^5 of each type, but the 40 straight flushes must be subtracted.)

Finally, consider hands containing a single pair but nothing better. There are $13(12 \cdot 11 \cdot 10)/6$ types of one pair hands, each occurring in $C_{4,2} \cdot 4 \cdot 4 \cdot 4$ ways; so

$$p(\text{one pair, no better}) = \frac{C_{4,2} \cdot 4^3 \cdot 13 \cdot 12 \cdot 11 \cdot 10/6}{C_{52,5}} = \frac{1,098,240}{C_{52,5}}$$

$$= .4226 \quad \left[\approx \frac{1}{2.4} \right].$$

Table 9 gives from best to worst a complete list of disjoint 5 card poker hands with the number of ways they can occur and their probabilities.

TABLE 9

5 Card Poker Hands and Their Probabilities

Hand	Number of possible ways	Probability (in 5 cards)
Straight flush	40	.000015
Four of a kind	624	.000240
Full house	3744	.001441
Flush	5108	.001965
Straight	10,200	.003925
Three of a kind	54,912	.021129
Two pair	123, 552	.047539
One pair	1,098,240	.422569
Worse than one pair	1,302,540	.501177
TOTALS	$2,598,960 = C_{52,5}$	1.000000

The three events we have not dealt with (3 of a kind, 2 pair, no pair) are the subject of Exercise 4.3.

Table 9 shows forcefully the pecking order of poker hands for 5 card stud, though it provides no justification for using the same order in draw games, 7 card stud, or wild-card games. Indeed, the main justification in these games is not probability but simplicity.

A poker player seldom uses in quantitative fashion the computed probabilities for being dealt a given 5 card hand. Much more important to a player is knowing the probability of obtaining a certain type of hand when part of it is already dealt. With this knowledge and observation of the other players and their exposed cards, the player can make an assessment of the probability of winning the deal by achieving the desired hand. We illustrate with a few sample hands.

Hand 1. In 5 card draw against one opponent you are dealt Queen, 7, 3, 2 of Diamonds and the King of Spades. Your probability of drawing one card and obtaining your flush is clearly $9/47 \approx .191$. If prior experience told you that your opponent was drawing 1 card to his two pair, then p(his full house) $= 4/47$. Your probability of winning this particular deal would then be $(9/47)(43/47) \approx .175$.

Hand 2. In 7 card stud you have 4 Spades among your first 5 cards. There are 12 other cards face up among which 2 are Spades. What is the probability of obtaining your flush on your next two cards? There are 7 Spades still hidden from you among the 35 unseen cards. The probability (at this moment) of getting *no* Spade in your next 2 cards is $(28/35)(27/34) \approx .635$. Hence

$$p(\text{completing your flush}) \approx 1 - .635 = .365.$$

Hand 3. The situation is almost the same as in hand 2 except that you have 3 Spades among your first 4 cards. Again there are 12 other cards face up, two of which are Spades. Now,

$$p(\text{no Spade among next 3 cards}) = \frac{28}{36} \cdot \frac{27}{35} \cdot \frac{26}{34} \approx .459$$

$$p(1 \text{ Spade among next 3 cards}) =$$

$$\underbrace{\frac{28}{36} \cdot \frac{27}{35} \cdot \frac{8}{34}}_{\text{Spade on 3rd}} + \underbrace{\frac{28}{36} \cdot \frac{8}{35} \cdot \frac{27}{34}}_{\text{Spade on 2nd}} + \underbrace{\frac{8}{36} \cdot \frac{28}{35} \cdot \frac{27}{34}}_{\text{Spade on 1st}}$$

$$= C_{3,1} \cdot \frac{8}{36} \cdot \frac{28}{35} \cdot \frac{27}{34} \approx .424$$

(the $C_{3,1}$ factor arises since there are 3 rounds in which the one Spade can arrive). Thus

$$p(\text{completing your flush}) \approx 1 - (.459 + .424) = .117.$$

This could just as easily have been done directly by adding $p(2$ more Spades) and $p(3$ more Spades).

Hand 4. You are dealt two pair (Jacks and 4's) in 5 card draw. If you draw 1 card, what is p(full house)? If you keep only the Jacks and draw three cards, what is p(3 Jacks or better)? Clearly

$$p(\text{draw 1 for full house}) = 4/47 \approx .0851 \qquad (\text{about 1 chance in 12}).$$

If three cards are drawn, we challenge the reader to verify that

$$p(\text{3 Jacks or Jacks up}^\dagger \text{ full house}) = C_{3,1} \cdot \frac{2}{47} \cdot \frac{45}{46} \cdot \frac{44}{45} \approx .1221$$

$$p(\text{4 Jacks}) = C_{3,2} \cdot \frac{2}{47} \cdot \frac{1}{46} \cdot \frac{44}{45} \approx .0027$$

$$p(\text{something over Jacks}^\ddagger \text{ full house}) = \frac{41}{47} \cdot \frac{3}{46} \cdot \frac{2}{45} \approx .0025.$$

Since the disjoint events described are the only ways to get three Jacks or better,

$$p(\text{draw 3 for three Jacks or better}) \approx .1221 + .0027 + .0025 = .1273,$$

or about 1 chance in 8. Thus *if* three Jacks is almost certain to win and two pair to lose, you are considerably better off drawing 3 cards and throwing away your low pair.

Having illustrated, we hope, the value of organized thinking, careful counting, and combinations, we leave these probability calculations to consider the more vital and more elusive question of whether and how much to bet.

Betting in poker: A simple model

When it is your bet in poker you must make a decision to *drop* (throw in your hand), *call* (pay the requested amount), or *raise* (increase the requested amount by some additional amount). As we shall now illustrate, a good poker player should base this decision upon a variety of

---

†In poker parlance, this means a full house containing three Jacks.
‡I.e. a full house in which the pair consists of Jacks.

quantitative estimates and projections (plus psychological and intuitive factors which we do not try to consider here). We will treat the special case of the final round of betting after all cards have been dealt. The simple model we construct takes into account the following often highly speculative information:

p = estimated probability of winning the pot.
c = current size of pot prior to present round of betting.
a = amount that must be paid to *call* all subsequent bets or raises.
r = projected size of raise made by the bettor.
n = estimated number of players involved throughout the final round of betting.

We assume (somewhat questionably) that n does not depend on r, so that (as in many small stake games) bluffing will be to no avail in the final round.

The bettor's decision on the last round whether to drop, call, or raise r units can be based upon the following expectation calculations. In determining payoffs (gains and losses) we base our estimates on the amount of money in the bettor's possession at the moment, not what he had at the start of the hand (we justify this a bit later). We compute

$$X(\text{drop}) = 0 \; [\text{he neither adds to nor subtracts from current holdings}]$$

$$X(\text{call}) = p(\underbrace{c + (n-1)a}_{\text{gain}}) + (1-p)\underbrace{(-a)}_{\text{loss}} = p(c + na) - a$$

$$X(\text{raise}) = p(c + (n-1)(a+r)) + (1-p)(-a-r)$$

$$= p(c + n(a+r)) - (a+r).$$

Comparing these expectations, we get

$$X(\text{raise}) > X(\text{call}) \Leftrightarrow X(\text{raise}) - X(\text{call}) > 0$$

$$\Leftrightarrow pnr - r > 0$$

$$\Leftrightarrow p > \frac{1}{n}.$$

Furthermore,

$$p > \frac{1}{n} \Rightarrow p > \frac{a}{na} \Rightarrow p > \frac{a}{c + na} \Leftrightarrow X(\text{call}) > X(\text{drop}).$$

We conclude that whenever $p > 1/n$, $X(\text{raise}) > X(\text{call}) > X(\text{drop})$, so a raise is in order. It can further be seen that, under our assumptions, the bettor should make r, the amount of his raise, the maximum limit allowed to maximize his expectation. Finally, if $p \leqslant 1/n$ then he should drop *unless* $p > a/(c + na)$, in which case call. Noting that $a/(c + na)$ is simply the ratio of what he must bet to the total final pot size, we can summarize our findings:

If $p > 1/n$, *raise* the limit. Otherwise,

$$\textit{call if} \qquad \frac{\text{cost of calling}}{\text{final pot size}} \leqslant p;$$

drop if this ratio exceeds p.

As a simple example, let $p = 1/6$, $c = \$20$, $n = 3$, and $a = \$5$. Since $p < 1/3$ the bettor should either call or drop. Since

$$p > \frac{5}{20 + 3 \cdot 5} = \frac{1}{7},$$

he should call. If c had been $\$10$, the model would suggest a drop (since $p < 5/25$). If p had been $1/2$, a raise should be made (regardless of c and a).

The model is clearly oversimplified. Its projections are not startling and smack of common sense (though the "last bet" behavior of many indicates that the model might be instructive—"but look at all the money I'd have won if the other 4 players were all bluffing"). We apply the model now to justify our claim above that past contributions to the pot are irrelevant to one's decision on how to bet at a later stage (though certainly the overall pot size c is highly relevant). Suppose the bettor has already contributed g units to the pot; let us compute expectations with the view that these g units are still his. Then

$$X(\text{drop}) = -g \qquad\qquad [\text{he lost } g \text{ units on the hand}]$$

$$X(\text{call}) = p(c + (n - 1)a - g) + (1 - p)(-a - g)$$

$$= p(c + na) - a - g$$

$$X(\text{raise}) = p(c + (n - 1)(a + r) - g) + (1 - p)(-a - r - g)$$

$$= p(c + n(a + r)) - (a + r) - g.$$

Comparing with our previous expectation calculation, we see that as we might have expected, each X value has been reduced by g, leaving their relative order unchanged. Thus the bettor's contribution to the pot so far has no bearing on his decision on the current bet. We have justified that fine and often violated poker (and gambling) maxim: "Don't throw good money after bad."

Our betting analysis is clearly only a beginning. We have ignored early round betting, the various types of poker (table stakes, pot limit, etc.), and the ubiquitous two-winner games. Perhaps more fundamentally, we have not discussed bluffing, reading the opponents, and numerous other subtleties that make poker difficult and inappropriate to analyze in a fully mathematical way. We trust nonetheless that our model and the probabilistic examples that preceded it will shed light on this most fascinating pastime.

Distributions in bridge

We consider only one minor aspect of the complex game of bridge—the problem of suit distribution probabilities and their implications for certain facets of the play. Our goal is to illustrate further the utility of combinations and careful counting in making rational decisions.

The number of different bridge hands was determined previously as $C_{52,13}$, a figure exceeding 600 billion. To compute the probability of a given suit distribution (like 4 of one suit and 3 in each of the others) in a random hand, we divide the number of bridge hands with such a distribution by $C_{52,13}$. We first consider the case of 5-4-3-1 distribution. If the specific suits (the 5 card suit, the 4 card suit, and the 3 card suit) are already determined, there are

$$C_{13,5} \cdot C_{13,4} \cdot C_{13,3} \cdot C_{13,1} = 1287 \cdot 715 \cdot 286 \cdot 13 = 3,421,322,190$$

such hands. There are, however, $P_{4,4} = 24$ ways of permuting the 4 *different sized* suits in a 5-4-3-1 distribution. Thus the probability of such a distribution is

$$p(5\text{-}4\text{-}3\text{-}1) = \frac{P_{4,4} \cdot C_{13,5} \cdot C_{13,4} \cdot C_{13,3} \cdot C_{13,1}}{C_{52,13}} \approx .129$$

(about one chance in eight).

As a second example consider the 4-3-3-3 hand. Here there are only $P_{4,1} = 4$ different suit arrangements (one for each choice of the 4 card suit). To get the total number of 4-3-3-3 hands we form the product $4 \cdot C_{13,4} \cdot (C_{13,3})^3$. Dividing by $C_{52,13}$ one gets the probability $p(4\text{-}3\text{-}3\text{-}3)$

$\approx .105$ (about one chance in 9 or 10). Similarly, the most likely distribution 4-4-3-2 has

$$p(4\text{-}4\text{-}3\text{-}2) = \frac{P_{4,2} \cdot (C_{13,4})^2 \cdot C_{13,3} \cdot C_{13,2}}{C_{52,13}} \approx .216 \text{ (about one hand in five)}.$$

The factor $P_{4,2} = 4 \cdot 3$ arises since there are 4 ways to choose the 3 card suit and then 3 ways to choose the 2 card suit. This uniquely determines the 4 card suits.

Fortified by these examples, we can give a general formula for the probability of any specific distribution $w\text{-}x\text{-}y\text{-}z$ (where clearly $w + x + y + z = 13$). Let n be the number of different *suit arrangements* in such a distribution. The only 3 possible values for n are

$$n = \begin{cases} 24 = P_{4,4} & \text{if all suits have different size} \\ 12 = P_{4,2} & \text{if exactly 2 suits have the same size} \\ 4 = P_{4,1} & \text{if 3 suits have the same size.} \end{cases}$$

Then, reasoning as illustrated in the cases considered previously, we get

$$p(w\text{-}x\text{-}y\text{-}z \text{ distribution}) = \frac{n \cdot C_{13,w} \cdot C_{13,x} \cdot C_{13,y} \cdot C_{13,z}}{C_{52,13}}.$$

As in the case of poker, the bridge player is not so much interested in advance probabilities of suit distributions as in distribution-related decisions which must be made during the play of the hand. Suppose you, as declarer, and your dummy partner have an 11 Spade trump suit between you, missing only the King and 2. Holding the Ace and Queen of Spades in your hand and possessing no special information from the bidding or play, you must decide whether to lead out your Ace (play for the *drop*) hoping the King will fall beneath it, or to lead up to your hand planning to play your Ace only if the player on your right (call him East) plays the King (play the *finesse*). How can you decide whether the drop or the finesse is the better percentage play? We analyze this in Table 10, which we now explain. We illustrate how the table is constructed by discussing line (3). If East holds the 2 but not the King of Spades, his hand then requires selecting 12 more cards from the 24 left unspecified (you and dummy have 26 and the K, 2 of Spades are accounted for). Hence there are $C_{24,12}$ ways to have this holding. Calculating the other entries in the "# of ways" column and observing that they *must* sum to $C_{26,13}$ (see Exercise 4.6), we find for example that the probability of Spade Holding (3) is

$$\frac{C_{24,12}}{C_{26,13}} = \frac{24!/(12!12!)}{26!/(13!13!)} = \frac{24!13!13!}{26!12!12!} = \frac{13\cdot13}{26\cdot25} = \frac{13}{50} = .26.$$

TABLE 10

Missing the K, 2

	East Spade holding	# of ways	Probability	Contribution to p(drop)	Contribution to p(finesse)
(1)	none	$C_{24,13}$	.24	0	0
(2)	K	$C_{24,12}$	.26	.26	.26
(3)	2	$C_{24,12}$	.26	.26	0
(4)	K2	$C_{24,11}$	.24	0	.24
	TOTAL	$C_{26,13}$	1.00	.52	.50

Finally, with East holding only the Spade 2, the drop will succeed (contributing .26 to p(drop)) and the finesse will fail (contributing 0 to p(finesse)). The analysis for the other holdings is similar. Note that for holding (2) the immediate appearance of East's King forces the correct play of the Ace even if the finesse had been planned. Consequently, .26 is placed in both the drop and finesse columns. We conclude from the overall totals that a 1-1 split has odds of 52:48 in its favor and that the drop play will work 52 percent of the time, with favorable odds of 52:50 over the finesse. The reader should see why p(drop) and p(finesse) need not add to 1. We now confess that for this particular problem there is a much more direct approach (see Exercise 4.8) and that our combinatorial method is just a warm-up for bigger and better things.

Consider now a similar situation where you and partner have 9 Spade trumps between you, missing only Q,4,3,2. You have the A,K,J,10,6 in your hand, enough entries to partner's hand, and no information on the opponents' holdings. The plan is first to lead out the Ace in the hopes of catching a singleton Queen. If this fails we again ask whether to play for the drop or to finesse. In Table 11, "x" represents a low card (in this case a 4, 3, or 2 of Spades). Note that in holdings (3) and (6) the Queen drops on the lead of the Ace; and in holding (8) the Ace reveals that East has *all* missing Spades, so the drop plan is abandoned in this case (but we chalk up a success in the drop column anyway). We conclude that the drop has favorable odds of about 58:56 and will work 57.8 percent of the time.

TABLE 11

Missing the Qxxx

	East Spade holding	# of ways	Probability	Contribution to p(drop)	Contribution to p(finesse)
(1)	none	$C_{22,13}$	.048	0	0
(2)	x	$3 \cdot C_{22,12}$	.187	0	0
(3)	Q	$C_{22,12}$	.062	.062	.062
(4)	xx	$3 \cdot C_{22,11}$	.203	.203	0
(5)	Qx	$3 \cdot C_{22,11}$	.203	.203	.203
(6)	xxx	$C_{22,10}$	.062	.062	.062
(7)	Qxx	$3 \cdot C_{22,10}$	.187	0	.187
(8)	Qxxx	$C_{22,9}$	.048	.048	.048
	TOTAL	$C_{26,13}$	1.000	.578	.562

Our final example shows that the finesse is sometimes the right tactic The situation is close to that of Table 11, except that you and partner have only 8 Spades (you still have A,K,J,10,6), missing Q,5,4,3,2. Again you lead out your Ace fishing for a singleton and then plan to face the drop or finesse question. Here we go again! Notice the appearance of the combination factors (e.g., in holding (4) there are $C_{4,2} = 6$ ways for East to have just 2 low Spades) and the beautiful symmetry of the third and fourth columns of the table. Here the odds strongly favor the finesse (about 51 : 33), which will succeed 50.8 percent of the time. A comparison of the results of Tables 11 and 12 provides a justification

TABLE 12

Missing the Qxxxx

	East Spade holding	# of ways	Probability	Contribution to p(drop)	Contribution to p(finesse)
(1)	none	$C_{21,13}$	.020	0	0
(2)	x	$C_{4,1} \cdot C_{21,12}$	.113	0	0
(3)	Q	$C_{21,12}$	.028	.028	.028
(4)	xx	$C_{4,2} \cdot C_{21,11}$	.203	0	0
(5)	Qx	$C_{4,1} \cdot C_{21,11}$	.136	.136	.136
(6)	xxx	$C_{4,3} \cdot C_{21,10}$	.136	.136	0
(7)	Qxx	$C_{4,2} \cdot C_{21,10}$	.203	0	.203
(8)	xxxx	$C_{21,9}$	.028	.028	.028
(9)	Qxxx	$C_{4,3} \cdot C_{21,9}$	.113	0	.113
(10)	Qxxxx	$C_{21,8}$	.020	0	0
	TOTAL	$C_{26,13}$	1.000	.328	.508

for the bridge maxim "eight ever (finesse), nine never." Nowhere do we justify mathematically that other old standby "one peek is worth two finesses."

By a similar reasoning the percentages on any drop vs. finesse question can be analyzed. Of course such computational activity would be frowned upon at the bridge table, but the methods are another illustration of the power of counting. We refer the interested reader to Oswald Jacoby's *On Gambling* which contains extensive tables for these situations and many others. While there are many other aspects of bridge that can be treated mathematically, we content ourselves with the above; but see Exercise 4.7 for some mathematics of game and slam bidding.

Keno type games

There is a variety of games in which a player selects or is given a set of numbers, some or all of which he hopes to match with numbers drawn without replacement from a larger set of numbers. The most widely known example of this process is probably Bingo, with which we deal briefly in Exercise 4.9. There is an immense number of variations on this theme, mostly originating from the English traveling carnival circuit of the late 19th century. We refer to these as Keno type games and we focus our attention here on the version of Keno which is played in casinos in Las Vegas and elsewhere.

In Casino Keno (or Race Horse Keno as it is sometimes called) a player receives a card with the numbers from 1 to 80 on it. He then marks the numbers he wants to play (anywhere from 1 to 15 of them) and indicates the amount of his bet. Twenty numbers are then drawn without replacement from a ping pong ball blower or in some other presumably random fashion. If an appropriate proportion of the marked numbers are drawn, the player gets a payoff somewhat less than that dictated by the true odds of what transpires.

About 80 percent of the play in Casino Keno is based on marking 10 numbers on the player's card, which is called a 10-spot ticket. Accordingly, we give a fairly thorough analysis of the 10-spot Keno probabilities and payoffs, leaving a few other questions to the exercises. Table 13 gives probabilities and payoffs for 10-spot Keno. We explain below how these probabilities are determined.

To explain how this table is constructed we first note that there are

$$C_{80,10} = 1,646,492,100,120$$

different ways to mark a 10-spot card. This will always be our denominator in the probability calculations. With 10 specific numbers marked and 20 random numbers drawn, there are $C_{20,10}$ (out of the $C_{80,10}$) ways for all 10 marked numbers to appear among the 20 numbers drawn. Hence

$$p(\text{all 10 marked numbers drawn}) = \frac{\text{\# ways it can happen}}{\text{\# of elementary events}}$$

$$= \frac{C_{20,10}}{C_{80,10}} \approx .00000001 .$$

TABLE 13

10 Spot Keno

# of marked Numbers drawn	# of ways	Probability $\left(\dfrac{\text{\# ways}}{C_{80,10}} \right)$	House payoff on a \$1 bet
10	$C_{20,10}$	.0000001	9,999
9	$C_{20,9} \cdot C_{60,1}$	.0000061	2,599
8	$C_{20,8} \cdot C_{60,2}$	.0001354	1,299
7	$C_{20,7} \cdot C_{60,3}$	.0016111	179
6	$C_{20,6} \cdot C_{60,4}$	.0114793	17
5	$C_{20,5} \cdot C_{60,5}$	.0514276	1
TOTAL	106,461,978,304	.0646596	

To see how products of combinations arise, consider the case of marking exactly 6 winning numbers. Assume, for our analysis, that the 20 random numbers have been determined, but remain unknown to the player. In how many ways can exactly 6 out of 10 marked numbers appear among the 20 numbers drawn? (In actuality, the player marks his ticket before the numbers are drawn, but our reinterpretation will not affect the results.) There are $C_{20,6}$ combinations of 6 marked numbers appearing among the 20 numbers drawn, but *for each such combination* there remain 4 marked numbers to be chosen among the 60 *undrawn* numbers (possible in $C_{60,4}$ ways). Thus there is a total of $C_{20,6} \cdot C_{60,4}$ equally likely ways to have exactly 6 of the 10 marked numbers drawn. The remainder of the table (except for the payoffs, which may vary slightly and are not up to us to compute!) should now be understandable and provides us with another delightful application of the power of combinatorial counting.

As can be seen from Table 13, the payoffs for each row are considerably below what the probabilities would dictate in a fair game situation. You will collect a payoff with probability .065 (about 1 time in 15) and the probability of losing your dollar is $1 - .065 = .935$ (if 4 or fewer of your marked numbers are drawn, you lose). Multiplying each payoff by its probability and adding, we obtain the expectation on a $1 bet:

$$X(10\text{-spot Keno}) \approx .001 + .016 + .175 + .288 + .195 + .051 - .935$$

$$\approx -.21.$$

We conclude by observing that the house edge on a 10-spot bet is close to 21 percent (you expect to lose 21 cents for each dollar bet). Exercise 4.11 illustrates the fact that all other Casino Keno bets have a similar house edge—making craps and even roulette and chuck-a-luck games seem like a gambler's haven in comparison.

Exercises

4.1   Consider a baseball team with a 25 player roster including 10 pitchers. Exactly one pitcher must be in the lineup, batting in the ninth position. Write down an expression for the number of possible batting orders.

4.2   A group of 9 people decide to split up into a bridge foursome and a backgammon twosome, with the remaining 3 people unoccupied.

a) Regarding two bridge foursomes as identical regardless of how partners are arranged, in how many different ways can these 9 people split up?

b) Now treating bridge foursomes as identical only if, for each player, partner and right hand opponent remain fixed, answer the question asked in a).

4.3   a) Verify the claim made in Table 9 about the number of possible 3 of a kind poker hands.

b) Do the same for the number of possible two pair poker hands.

c) If all the information in Table 9 has been obtained except for the "worse than one pair" row, explain in words how that last information can be deduced. Do not make the actual computation.

4.4   Write expressions for the number of possible bridge hands with each of the following distributions.

a) 10-1-1-1.

b) 5-5-3-0.

c) 7-3-2-1.

4.5   a) Write an expression for the number of possible bridge *deals* (a deal is the passing out of a separate hand to each of the 4 players).

b) Use a) to write an expression for the probability of a bridge deal in which *each* player is dealt a 13 card suit. This event seems to be reported as happening somewhere almost every year. Does your answer suggest that honest reporting and truly random dealing are taking place?

4.6 Reread the discussion following Table 10, where it is observed that the "# of Ways" column must sum to $C_{26,13}$. Prove directly from the definition of combinations that indeed $C_{24,13} + C_{24,12} + C_{24,12} + C_{24,11} = C_{26,13}$. Hint: give all the fractions involved a common denominator of $13! \cdot 13!$.

4.7 (Game and slam bidding) The following table summarizes points awarded (or lost) in duplicate bridge scoring in various nonvulnerable and vulnerable successes (and failures) involving *no-trump* bids. A *game* bid of 3NT requires 9 tricks, a *small slam* (6NT) requires 12 tricks, and a *grand slam* (7NT) requires all 13 tricks. *Down one* means 1 less trick was made than was bid.

	Game	Small slam	Grand slam	Down one	Bid 2NT and make 8(9) tricks	Bid 5NT and make 11(12) tricks
non-vulnerable	400	990	1520	-50	120 (150)	460 (490)
vulnerable	600	1440	2220	-100	120 (150)	660 (690)

Assume that contemplated bids will never be off by more than 1 trick and ignore the possibility of an opponent's double. Also assume your goal is to maximize points scored.

a) Let p = estimated probability of making a no-trump game (9 tricks). Show that if you are nonvulnerable then

$$X(\text{Bid 3NT}) = p(400) + (1-p)(-50)$$

and

$$X(\text{Bid 2NT}) = p(150) + (1-p)(120).$$

Conclude that under our assumptions a non-vulnerable no-trump game should be bid when $p > 17/42$ (about .4), and bidding should stop at 2NT when $p < 17/42$. Then carry out a similar analysis in the vulnerable case and show that the borderline value of p is $22/67$ (about $1/3$).

b) Let p = estimated probability of making a small slam (12 tricks) in no-trump. Using the ideas of a), show that the borderline probability beyond which at least 6NT should be bid is $p = 51/101$ when non-vulnerable and $p = 76/151$ when vulnerable. Thus p is very close to $1/2$ in each case.

c) Let p = estimated probability of making a grand slam (13 tricks) in no-trump. Also, note that making all 13 tricks when a small slam is bid adds 30 points (1 overtrick) to the "small slam" column of our table. Show that the borderline probabilities for the grand slam bid are $p = 52/77$

(non-vulnerable) and $p = 154/229$ (vulnerable). Thus p is very close to 2/3 in each case.

4.8   a) Reestablish the conclusions of Table 10 by completing the following simple line of argument: with only 2 Spades among the opponents' 26 cards, consider the hand containing the Spade Two. There are 12 slots remaining in that hand while there are 13 slots in the other opponent's hand in which the Spade King can be found....

b) The argument in a) shows directly that the odds in favor of a 1-1 split are $13:12$. Discuss and try to resolve the following "paradox." To improve the odds of the 1-1 split, first lead out one card in each side suit (besides Spades). Then opponents have a total of 20 cards between them and when the argument of a) is applied the odds in favor of a 1-1 split have been improved to $10:9$. Therefore you improve your chances for the drop by playing out side suits first!

4.9   A Bingo card has 5 columns (B-I-N-G-O) with 5 numbers in each column. All numbers are distinct and values range from 1 to 75 (in England from 1 to 90). For the moment we ignore the "free" square usually found in the center of the card.

a) In the first five draws from the 75 numbers in the Bingo drum how many different sets of 5 numbers can be drawn?

b) By counting the number of ways to win on a particular Bingo card (rows, columns, diagonals, but not "four corners"), compute the probability of winning with a single card after just 5 numbers have been drawn.

c) Now answer b) again, this time allowing for a "free" square and a win using "four corners." Hint: on a given card there are 5 wins requiring only four numbers, and each of these wins can occur on 71 different draws of 5 numbers.

4.10   a) Consider a 1-spot Keno bet in which you mark only 1 number and hope it will be among the 20 (out of a possible 80) numbers drawn. If your number is drawn you win $2.20 and otherwise you lose your $1 bet. Show that the house edge in 1-spot Keno is exactly 20 percent.

b) A 2-spot Keno bet wins $12 if both marked numbers are drawn, with the $1 stake being lost otherwise. Prove that the house edge for this bet is $69/316 \approx 21.84$ percent.

4.11   a) Consider a general n-spot Keno ticket where n numbers are marked and then 20 out of 80 numbers are drawn. Prove that for $k \leqslant n$,

$$p(\text{exactly } k \text{ marked numbers will be drawn}) = \frac{C_{20,k} \cdot C_{60,n-k}}{C_{80,n}}.$$

b) Argue that the house payoff on a perfect 10-spot Keno ticket (all 10 numbers are drawn) could be raised from $9,999 to $99,999 without significantly lowering the house edge. Why don't all casinos make this change as a means of attracting publicity and customers?

## CHAPTER 5

Play it Again Sam: The Binomial Distribution

Games and repeated trials

The expectation concept provides a way of predicting what should happen, on the average, if one plays certain games long enough. Yet there are unquestionably people who have made negative expectation bets a large number of times and ended up with a profit. In this chapter we consider the mathematics of repeated plays (trials) of a given game (experiment). The technique developed will enable us to make theoretical estimates of just how likely one is to "beat the odds" in a specific number of bets at given odds. The relevance of such a theory should be apparent to the reader, for almost all gambling is of this repeated play variety. We comment further that any repeated activity whose probability of success remains fixed on each repetition will be subjected to the mathematics developed. To stretch a point, we might argue that an individual basketball player (or a whole team) shooting 40 shots (repeated trials) a game is merely acting out our theory, and that the surprise wins, routine performances, and upset losses are natural consequences of the theory's laws. Such dehumanization is, to be sure, highly oversimplified and often innaccurate, but it does have some validity and might even provide solace in difficult times.

The binomial distribution

We motivate the general case with a question raised in passing by Dostoyevsky's *The Gambler* (Chapter 1) and partially treated in *Question* 6 of Chapter 4. Given a European roulette wheel (37 numbers including a 0 but no 00), what is the probability of getting exactly 3 zeros in 10 spins of the wheel? A typical sequence of 10 spins in which 3 zeros occur might be $XX0XXX00XX$, where X represents "not 0." The probability of this particular sequence of independent spins is

$$\frac{36}{37} \cdot \frac{36}{37} \cdot \frac{1}{37} \cdot \frac{36}{37} \cdot \frac{36}{37} \cdot \frac{36}{37} \cdot \frac{1}{37} \cdot \frac{1}{37} \cdot \frac{36}{37} \cdot \frac{36}{37} = \left(\frac{36}{37}\right)^7\left(\frac{1}{37}\right)^3 \approx .000016.$$

To obtain the desired overall probability, we need to multiply this probability by the number of disjoint, equally likely ways 3 zeros can occur. This is a "combination" question with answer $C_{10,3}$. We conclude that

$$p(3 \text{ zeros in } 10 \text{ spins}) = C_{10,3}\left(\frac{36}{37}\right)^7\left(\frac{1}{37}\right)^3 \approx 120 \cdot .000016 = .00192$$

or about 1 chance in 500. The reader should now see that the reasoning applied above generalizes beautifully to the case where the probability of success in each *independent trial* is p (rather than $1/37$), the probability of failure is $q = 1 - p$ (rather than $36/37$), the number of trials is n (rather than 10), and the specified number of successes is r (rather than 3). We give one more example before presenting the general result.

You make 100 bets of \$1 on "pass" in casino craps. What is your probability of winning exactly \$6? Recall that the probability p of success in a pass bet is $p = .493$, so that $q = .507$. The only way you can win exactly \$6 in 100 even payoff \$1 bets is to win 53 and lose 47. Hence, in the $n = 100$ trials, we are interested in precisely $r = 53$ successes and $n - r = 47$ failures. Reasoning as before,

$$p(\text{winning } \$6) = C_{100,53} \cdot (.507)^{47}(.493)^{53} = C_{n,r}q^{n-r}p^r.$$

We now have ample justification for stating an important result in statistics which has considerable utility in the theory of gambling.

If an experiment with fixed probability p of success is repeated for n independent trials, then

$$p(\text{exactly } r \text{ successes in } n \text{ trials}) = C_{n,r}q^{n-r}p^r \quad (q = 1 - p).$$

This distribution of probabilities for $r = 0, 1, 2, \ldots, n$ successes is called the *binomial distribution*, and we present below a few of its many types of applications. We caution that the hypotheses of fixed probability and independent trials must be satisfied before the binomial distribution results can be applied.

Application 1. An honest coin is flipped 8 times. What is the probability of getting exactly r heads? Since $p = q = 1/2$ and $q^{8-r}p^r = (1/2)^8 = 1/256$, the solution is given by the following table:

# of Heads (r)	# of Ways	Probability
0	$C_{8,0} = 1$	1/256
1	$C_{8,1} = 8$	8/256
2	$C_{8,2} = 28$	28/256
3	$C_{8,3} = 56$	56/256
4	$C_{8,4} = 70$	70/256
5	$C_{8,5} = 56$	56/256
6	$C_{8,6} = 28$	28/256
7	$C_{8,7} = 8$	8/256
8	$C_{8,8} = 1$	1/256

The symmetry of the probabilities in the table results from the fact that $p = 1/2$. Note that $p(4 \text{ heads}) = 70/256 = .273$. Likewise,

$$p(7 \text{ or more heads}) = C_{8,7} \cdot \frac{1}{256} + C_{8,8} \cdot \frac{1}{256} = \frac{9}{256}$$

and

$$p(6 \text{ or fewer heads}) = 1 - p(7 \text{ or more heads}) = \frac{247}{256}.$$

Application 2.   An honest die is rolled 12 times. Compute the probability that a *five or six* will turn up 4 or more times. Here "success" is the event 5 or 6, so $p = 1/3$ and $q = 2/3$. Thus

$p(4 \text{ or more successes})$

$$= 1 - p(3 \text{ or fewer successes})$$

$$= 1 - [\, p(3 \text{ successes}) + p(2 \text{ successes})$$

$$+ p(1 \text{ success}) + p(0 \text{ successes})\,]$$

$$= 1 - \left[C_{12,3}\left(\frac{2}{3}\right)^9\left(\frac{1}{3}\right)^3 + C_{12,2}\left(\frac{2}{3}\right)^{10}\left(\frac{1}{3}\right)^2 \right.$$

$$\left. + C_{12,1}\left(\frac{2}{3}\right)^{11}\left(\frac{1}{3}\right)^1 + C_{12,0}\left(\frac{2}{3}\right)^{12}\left(\frac{1}{3}\right)^0 \right]$$

$$= 1 - \left[220 \cdot \frac{512}{3^{12}} + 66 \cdot \frac{1024}{3^{12}} + 12 \cdot \frac{2048}{3^{12}} + 1 \cdot \frac{4096}{3^{12}} \right]$$

$$= 1 - \frac{208{,}896}{531{,}441} \approx .607.$$

Application 3. Assume that the Phillies and the Yankees are in the world series, that the Phillies have a 3/5 chance of winning any given game, and that the games are independent experiments. What is the probability of a 7 game series? A 7 game series will occur when and only when each team wins 3 of the first 6 games. Thus

$$p(\text{Phillies win 3 out of 6}) = C_{6,3}(.4)^3(.6)^3 = 20 \cdot .064 \cdot .216 = .276.$$

The binomial distribution, clearly tied in with probabilities and combinations, is intimately linked to a familiar result of high school algebra and to Blaise Pascal's famous triangle. We briefly state the connection here to make these beautiful ideas easier to remember and appreciate. Using the familiar binomial expansions

$$(a + b)^2 = 1a^2 + 2ab + 1b^2; \qquad (a + b)^3 = 1a^3 + 3a^2b + 3ab^2 + 1b^3$$

as motivation and counting arguments similar to those previously considered, we are led to

The Binomial Theorem. For *any numbers* q *and* p *and any positive integer* n,

$$(q + p)^n = C_{n,0}q^n + C_{n,1}q^{n-1}p + \cdots + C_{n,n-1}qp^{n-1} + C_{n,n}p^n.$$

While we do not offer a formal proof [†] of this important result, we invite the reader to think about how

$$(q + p)^n = (q + p)(q + p)(q + p) \cdots (q + p)$$

can be expanded by successive multiplication. Each term in the expansion is formed by choosing either q or p from each of the n factors $(q + p)$. For each $r = 0, 1, 2, \ldots, n$ there are precisely $C_{n,r}$ distinct ways to choose r of the p's from the n factors $(q + p)$. This explains the appearance of the terms $C_{n,r}q^{n-r}p^r$ in the binomial theorem.

If p and q happen to be complementary probabilities ($q + p = 1$) with p the probability of success, then $(q + p)^n = 1$, so the binomial theorem expresses 1 as a sum of $n + 1$ terms. From our binomial distribution results, we see that these $n + 1$ terms can be interpreted respectively as the probability of $r = 0, 1, 2, \ldots, n - 1, n$ successes in n independent trials. Since these $n + 1$ disjoint events are the only outcomes possible, their probabilities must of course add up to 1.

[†] See I. Niven's *Mathematics of Choice*, NML vol. 15, MAA Washington (1965), p. 34.

This circle (or triangle) of ideas is nicely completed by observing that the aptly named *binomial coefficients* $C_{n,r}$ can be obtained not only from factorials and binomial expansions, but also by constructing *Pascal's triangle*.

r n	0	1	2	3	4		
0	1						
1	1	1					
2	1	2	1				
3	1	3	3	1			
4	1	4	6	4	1		
5	1	5	10	10	5	1	
6	1	6	(15)	20	15	6	1

The entries in each row (after row 0) are obtained by summing the two entries in the row above directly on top of and diagonally to the left of the desired entry. When a blank is encountered in this summing process it is taken to be 0. The circled entry 15 in row 6, column 2 is thus obtained as the sum of 5 and 10. It is a remarkable fact that the entry in row n and column r of this triangle is precisely $C_{n,r}$ (e.g., $15 = C_{6,2} = 6!/(2!4!)$)—for this to work we must start our row and column count at 0). The entries in row n are the respective coefficients in the expansion of $(q + p)^n$.

Beating the odds and the "law" of averages

It should be apparent that the binomial distribution gives us, at least in theory, a means of computing our chances of winning a specified amount (i.e., having r successes) in a specified number n of plays of a constant bet gambling game with constant probability p of success. It should be equally apparent, however, that working with realistic and therefore often large values of n and r leads to computations (for $C_{n,r}$ and high powers of p and q) which are lengthy enough to tax even a computer. To make things worse, we are usually interested not simply in the probability of winning (or losing) a specified amount, but *at least* that amount, in which case many such computations must be summed. In this section we state theoretical results which provide easily computed and (for large n) accurate approximations for the desired probabilities. We then use the method described to show how to construct exceedingly revealing tables concerning prospects for "beating the odds" over a long series of bets.

In Figure 4 we exhibit a series of frequency charts (or *histograms*) which reflect the probabilities of various profits and losses in repeated "red" bets on a Las Vegas roulette wheel ($p = 18/38 \approx .474$, stake = $1). The first 4 charts (4, 8, 16, and 32 spins) were constructed by computing binomial distribution probabilities, and are therefore exact. The 64 spin chart was obtained using the *normal approximation* which we now discuss. Before beginning this discussion we make some qualitative observations about the charts in Figure 4. Initially the charts peak

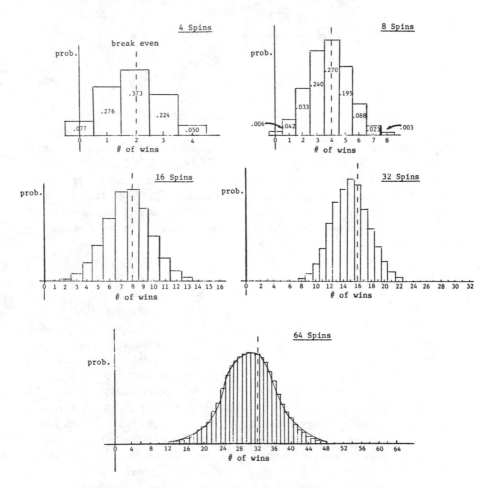

Figure 4   The normal approximation to the binomial distribution ($p = .474$).

around 0 profit (# of wins $r = n/2$) though some asymmetry in favor of losses is apparent. As the number of plays increases the peaks move leftward from the $r = n/2$ break-even point, but a bell-shaped symmetry about the peaks becomes very striking. In the final chart we superimpose a perfect bell-shaped *normal distribution* curve over the binomial distribution rectangles to emphasize this point. Finally, we note that since each rectangle in each chart has a base width of 1 unit, heights of the rectangles (probabilities) can be identified with the areas, which must, for each chart, add up to 1.

The bell-shaped normal distribution curve is obtained by graphing exponential functions of the form $ae^{-c(x-b)^2}$ where the constants a, b, c are chosen to ensure that

a)   The total area under the curve is 1.
b)   The peak of the curve is at an appropriate point b (it turns out to be the expected number np of successes in n trials) on the horizontal axis.
c)   The sharpness of the peak reflects the particular binomial situation being approximated.

The fundamental result which we now assert is:

The normal approximation to the binomial distribution: For cases in which $np > 5$ and $nq > 5$, an appropriate normal distribution will provide a "good" approximation to the binomial distribution.

Since the approximating exponential curve and the areas under various parts of it are no easier (in fact harder) to compute exactly than the corresponding binomial areas (probabilities), it would seem that we have gained very little computationally. The key point is, however, that every normal probability curve can be reduced to a *standard normal curve* (defined by the function $\left(1/\sqrt{2\pi}\right)e^{-z^2/2}$). Table 14 provides a way of computing areas under the standard normal curve. For each number Z on the horizontal axis, the table gives the area under the standard normal curve to the *right* of Z. Areas to the left of Z can then be obtained by subtracting from 1. Our task now is to explain how to compute areas in any binomial distribution chart (with $np > 5$ and $nq > 5$) by referring to Table 14.

Given a binomial distribution situation (with n and p specified), let r denote a given number of successes (in the n repeated independent trials).

Let

$$Z = \frac{r - \frac{1}{2} - np}{\sqrt{npq}}.$$

TABLE 14

Areas under the Standard Normal Curve

For each **Z**, the entry in the table is the
area under the curve to the right of **Z**.

z	0.00	0.01	0.02	0.03	0.04	0.05	0.06	0.07	0.08	0.09
3.4	0.0003	0.0003	0.0003	0.0003	0.0003	0.0003	0.0003	0.0003	0.0003	0.0002
3.3	0.0005	0.0005	0.0005	0.0004	0.0004	0.0004	0.0004	0.0004	0.0004	0.0003
3.2	0.0007	0.0007	0.0006	0.0006	0.0006	0.0006	0.0006	0.0005	0.0005	0.0005
3.1	0.0010	0.0009	0.0009	0.0009	0.0008	0.0008	0.0008	0.0008	0.0007	0.0007
3.0	0.0013	0.0013	0.0013	0.0012	0.0012	0.0011	0.0011	0.0011	0.0010	0.0010
2.9	0.0019	0.0018	0.0017	0.0017	0.0016	0.0016	0.0015	0.0015	0.0014	0.0014
2.8	0.0026	0.0025	0.0024	0.0023	0.0023	0.0022	0.0021	0.0021	0.0020	0.0019
2.7	0.0035	0.0034	0.0033	0.0032	0.0031	0.0030	0.0029	0.0028	0.0027	0.0026
2.6	0.0047	0.0045	0.0044	0.0043	0.0041	0.0040	0.0039	0.0038	0.0037	0.0036
2.5	0.0062	0.0060	0.0059	0.0057	0.0055	0.0054	0.0052	0.0051	0.0049	0.0048
2.4	0.0082	0.0080	0.0078	0.0075	0.0073	0.0071	0.0069	0.0068	0.0066	0.0064
2.3	0.0107	0.0104	0.0102	0.0099	0.0096	0.0094	0.0091	0.0089	0.0087	0.0084
2.2	0.0139	0.0136	0.0132	0.0129	0.0125	0.0122	0.0119	0.0116	0.0113	0.0110
2.1	0.0179	0.0174	0.0170	0.0166	0.0162	0.0158	0.0154	0.0150	0.0146	0.0143
2.0	0.0228	0.0222	0.0217	0.0212	0.0207	0.0202	0.0197	0.0192	0.0188	0.0183
1.9	0.0287	0.0281	0.0274	0.0268	0.0262	0.0256	0.0250	0.0244	0.0239	0.0233
1.8	0.0359	0.0352	0.0344	0.0336	0.0329	0.0322	0.0314	0.0307	0.0301	0.0294
1.7	0.0446	0.0436	0.0427	0.0418	0.0409	0.0401	0.0392	0.0384	0.0375	0.0367
1.6	0.0548	0.0537	0.0526	0.0516	0.0505	0.0495	0.0485	0.0475	0.0465	0.0455
1.5	0.0668	0.0655	0.0643	0.0630	0.0618	0.0606	0.0594	0.0582	0.0571	0.0559
1.4	0.0808	0.0793	0.0778	0.0764	0.0749	0.0735	0.0722	0.0708	0.0694	0.0681
1.3	0.0968	0.0951	0.0934	0.0918	0.0901	0.0885	0.0869	0.0853	0.0838	0.0823
1.2	0.1151	0.1131	0.1112	0.1093	0.1075	0.1056	0.1038	0.1020	0.1003	0.0985
1.1	0.1357	0.1335	0.1314	0.1292	0.1271	0.1251	0.1230	0.1210	0.1190	0.1170
1.0	0.1587	0.1562	0.1539	0.1515	0.1492	0.1469	0.1446	0.1423	0.1401	0.1379
0.9	0.1841	0.1814	0.1788	0.1762	0.1736	0.1711	0.1685	0.1660	0.1635	0.1611
0.8	0.2119	0.2090	0.2061	0.2033	0.2005	0.1977	0.1949	0.1922	0.1894	0.1867
0.7	0.2420	0.2389	0.2358	0.2327	0.2296	0.2266	0.2236	0.2206	0.2177	0.2148
0.6	0.2743	0.2709	0.2676	0.2643	0.2611	0.2578	0.2546	0.2514	0.2483	0.2451
0.5	0.3085	0.3050	0.3015	0.2981	0.2946	0.2912	0.2877	0.2843	0.2810	0.2776
0.4	0.3446	0.3409	0.3372	0.3336	0.3300	0.3264	0.3228	0.3192	0.3156	0.3121
0.3	0.3821	0.3783	0.3745	0.3707	0.3669	0.3632	0.3594	0.3557	0.3520	0.3483
0.2	0.4207	0.4168	0.4129	0.4090	0.4052	0.4013	0.3974	0.3936	0.3897	0.3859
0.1	0.4602	0.4562	0.4522	0.4483	0.4443	0.4404	0.4364	0.4325	0.4286	0.4247
0.0	0.5000	0.4960	0.4920	0.4880	0.4840	0.4801	0.4761	0.4721	0.4681	0.4641
−0.0	0.5000	0.5040	0.5080	0.5120	0.5160	0.5199	0.5239	0.5279	0.5319	0.5359
−0.1	0.5398	0.5438	0.5478	0.5517	0.5557	0.5596	0.5636	0.5675	0.5714	0.5753
−0.2	0.5793	0.5832	0.5871	0.5910	0.5948	0.5987	0.6026	0.6064	0.6103	0.6141
−0.3	0.6179	0.6217	0.6255	0.6293	0.6331	0.6368	0.6406	0.6443	0.6480	0.6517
−0.4	0.6554	0.6591	0.6628	0.6664	0.6700	0.6736	0.6772	0.6808	0.6844	0.6879
−0.5	0.6915	0.6950	0.6985	0.7019	0.7054	0.7088	0.7123	0.7157	0.7190	0.7224
−0.6	0.7257	0.7291	0.7324	0.7357	0.7389	0.7422	0.7454	0.7486	0.7517	0.7549
−0.7	0.7580	0.7611	0.7642	0.7673	0.7704	0.7734	0.7764	0.7794	0.7823	0.7852
−0.8	0.7881	0.7910	0.7939	0.7967	0.7995	0.8023	0.8051	0.8078	0.8106	0.8133
−0.9	0.8159	0.8186	0.8212	0.8238	0.8264	0.8289	0.8315	0.8340	0.8365	0.8389
−1.0	0.8413	0.8438	0.8461	0.8485	0.8508	0.8531	0.8554	0.8577	0.8599	0.8621
−1.1	0.8643	0.8665	0.8686	0.8708	0.8729	0.8749	0.8770	0.8790	0.8810	0.8830
−1.2	0.8849	0.8869	0.8888	0.8907	0.8925	0.8944	0.8962	0.8980	0.8997	0.9015
−1.3	0.9032	0.9049	0.9066	0.9082	0.9099	0.9115	0.9131	0.9147	0.9162	0.9177
−1.4	0.9192	0.9207	0.9222	0.9236	0.9251	0.9265	0.9278	0.9292	0.9306	0.9319
−1.5	0.9332	0.9345	0.9357	0.9370	0.9382	0.9394	0.9406	0.9418	0.9429	0.9441
−1.6	0.9452	0.9463	0.9474	0.9484	0.9495	0.9505	0.9515	0.9525	0.9535	0.9545
−1.7	0.9554	0.9564	0.9573	0.9582	0.9591	0.9599	0.9608	0.9616	0.9625	0.9633
−1.8	0.9641	0.9649	0.9656	0.9664	0.9671	0.9678	0.9686	0.9693	0.9699	0.9706
−1.9	0.9713	0.9719	0.9726	0.9732	0.9738	0.9744	0.9750	0.9756	0.9761	0.9767
−2.0	0.9772	0.9778	0.9783	0.9788	0.9793	0.9798	0.9803	0.9808	0.9812	0.9817
−2.1	0.9821	0.9826	0.9830	0.9834	0.9838	0.9842	0.9846	0.9850	0.9854	0.9857
−2.2	0.9861	0.9864	0.9868	0.9871	0.9875	0.9878	0.9881	0.9884	0.9887	0.9890
−2.3	0.9893	0.9896	0.9898	0.9901	0.9904	0.9906	0.9909	0.9911	0.9913	0.9916
−2.4	0.9918	0.9920	0.9922	0.9925	0.9927	0.9929	0.9931	0.9932	0.9934	0.9936
−2.5	0.9938	0.9940	0.9941	0.9943	0.9945	0.9946	0.9948	0.9949	0.9951	0.9952
−2.6	0.9953	0.9955	0.9956	0.9957	0.9959	0.9960	0.9961	0.9962	0.9963	0.9964
−2.7	0.9965	0.9966	0.9967	0.9968	0.9969	0.9970	0.9971	0.9972	0.9973	0.9974
−2.8	0.9974	0.9975	0.9976	0.9977	0.9977	0.9978	0.9979	0.9979	0.9980	0.9981
−2.9	0.9981	0.9982	0.9982	0.9983	0.9984	0.9984	0.9985	0.9985	0.9986	0.9986
−3.0	0.9987	0.9987	0.9987	0.9988	0.9988	0.9989	0.9989	0.9989	0.9990	0.9990
−3.1	0.9990	0.9991	0.9991	0.9991	0.9992	0.9992	0.9992	0.9992	0.9993	0.9993
−3.2	0.9993	0.9993	0.9994	0.9994	0.9994	0.9994	0.9994	0.9995	0.9995	0.9995
−3.3	0.9995	0.9995	0.9995	0.9996	0.9996	0.9996	0.9996	0.9996	0.9996	0.9997
−3.4	0.9997	0.9997	0.9997	0.9997	0.9997	0.9997	0.9997	0.9997	0.9997	0.9998

Then $p(r$ or more successes in n trials$) \approx$ the area (from Table 14) to the right of the computed Z value. A careful look at the way the rectangle corresponding to r wins in Figure 4 has a base with endpoints $r - 1/2$ and $r + 1/2$ should strongly suggest where the $-1/2$ comes from in the formula for Z. By converting the number r of successes to profit (or loss), we are now able to answer the probability questions in this binomial situation about repeated, fixed odds bets. We always assume a \$1 bet, but it should be clear how to deal with other size bets (see Exercise 5.4). We illustrate these ideas by first answering some questions based upon the situation treated in Figure 4 ($p = .474$) and an even money house payoff.

1. What is p(ending up ahead after 64 bets of \$1)? We are interested in $r = 33$, since 33 or more wins is equivalent to a profit. Hence

$$Z = \frac{32.5 - 64 \cdot .474}{\sqrt{64 \cdot .474 \cdot .526}} = \frac{2.164}{3.99} \approx .54.$$

Consulting Table 14, we see that the area corresponding to a Z value of .54 is .2946, so p(ending up ahead) $\approx .295$.

2. What is p(winning exactly \$20 after 64 bets of \$1)? The desired area cannot be obtained directly from Table 14; but, reasoning geometrically, this area will be the difference between the areas generated for $r = 42$ (win \$20) and $r = 43$ (win \$22). Since

$$Z_{42} = \frac{41.5 - 64 \cdot .474}{3.99} = 2.795, \quad \text{and} \quad Z_{43} = \frac{42.5 - 64 \cdot .474}{3.99} = 3.045,$$

we conclude that p(winning exactly \$20) $\approx .0026 - .0012 = .0014$.

3. What is p(losing \$10 or more after 64 bets of \$1)? Since losing \$10 is the same as *not* winning 28 or more bets (check this),

$$p(\text{losing } \$10 \text{ or more}) = 1 - p(r \geqslant 28).$$

Here $Z_{28} = -.71$, so $p(r \geqslant 28) \approx .761$. Finally,

$$p(\text{losing } \$10 \text{ or more}) \approx 1 - .761 = .239.$$

4. What is p(being even or ahead after 10,000 bets)? Here $r = 5,000$, so

$$Z = \frac{4999.5 - 10,000 \cdot .474}{\sqrt{10,000 \cdot .474 \cdot .526}} \approx 5.20.$$

This Z value can be seen from the table to correspond to an area of essentially 0 (certainly well below .0002 at any rate). Accordingly, p(being even or ahead) ≈ 0, so with such a large number of bets we have no right to expect anything but an overall loss.

5.   What is p(winning \$40 or more) in 500 roulette bets of \$1 on a single number? If your number (paying 35 to 1) comes up r times in 500 spins, your profit is $r \cdot 35 + (500 - r)(-1)$. To obtain the number of wins needed for a \$40 profit, we solve the equation

$$r \cdot 35 + (500 - r)(-1) = 40$$

for r to obtain $36r = 540$ or $r = 15$. Since for each bet the probability of winning is $p = 1/38 \approx .026$, we obtain

$$Z = \frac{14.5 - 500 \cdot .026}{\sqrt{500 \cdot .026 \cdot .974}} = \frac{1.5}{3.56} \approx .42.$$

We conclude that p(winning \$40 or more) $\approx .337$.

The above reasoning should convince the reader of the power, economy, and applicability of the normal approximation. Using this reasoning extensively, we can construct the highly informative Table 15. One can assess from this table one's prospects at various probabilities, house odds, and numbers of repeated plays. Some of the particular probabilities and payoffs have been chosen to correspond to well known gambling house bets. It should be pointed out that our analysis (and each row of Table 15) can only deal with a particular type of bet in a particular game. In games such as craps or roulette where a variety of bets are available, each type of bet must be analyzed separately. Activities such as slot machines and racetrack betting with fixed house edge but a variety of unpredictable payoffs would be very hard to describe in the form of Table 15.

The implications of Table 15 should be sobering to the devoted casino gambler. At any number of repeated plays of a positive house edge game, there is a possibility of winning and even winning big. But as the number of plays increases, the probability of winning drops, and the drop is more dramatic for larger house edge games (look at the $D = 0$ columns). The results *do not* say that a winning streak will tend to be counterbalanced by a corresponding run of losses (or vice versa!). They do say that anyone who expects to gamble at unfavorable odds for any length of time should hold little hope for turning a profit, and the longer the time and the higher the house edge, the less this hope should be. This is a consequence of the "law of averages," and this law is one

TABLE 15

Fighting the Odds on Repeated $1 Constant Odds Bets

Game-House Payoff	P(Winning Single Bet)	House Edge	# of Bets	p(Ending Ahead $D or More)				p(Ending Behind $D or More)			
				D = 30	D = 20	D = 10	D = 0	D = 0	D = 10	D = 20	D = 30
Fair Game – Even Money	.5	0%	50	.000	.003	.101	.556	.556	.101	.003	.000
			100	.002	.028	.184	.540	.540	.184	.028	.002
			500	.097	.198	.344	.518	.518	.344	.198	.097
Fair Game – 4 to 1	.2	0%	50	.026	.108	.298	.570	.570	.298	.108	.026
			100	.085	.191	.354	.550	.550	.354	.191	.085
			500	.269	.348	.433	.522	.522	.433	.348	.269
Craps – Even Money	.493	1.4%	50	.000	.002	.085	.516	.596	.120	.004	.000
			100	.000	.020	.149	.483	.595	.224	.039	.003
			500	.053	.122	.236	.393	.641	.466	.297	.163
Roulette – Even Money	.474	5.2%	50	.000	.001	.050	.408	.607	.184	.010	.000
			100	.000	.008	.077	.334	.734	.355	.084	.008
			500	.007	.021	.057	.128	.889	.781	.629	.452
Roulette – 8 to 1	.105	5.2%	50	.067	.150	.282	.454	.546	.361	.210	.102
			100	.096	.164	.257	.373	.627	.500	.257	.164
			500	.191	.233	.280	.331	.669	.615	.558	.500
Bookmaker – Even Money	.45	10%	50	.000	.000	.024	.284	.803	.286	.022	.000
			100	.000	.002	.028	.183	.865	.541	.183	.027
			500	.000	.001	.004	.014	.989	.967	.918	.827
Bookmaker – Even Money	.40	20%	50	.000	.000	.003	.098	.943	.561	.096	.002
			100	.000	.000	.002	.027	.983	.869	.543	.179
			500	.000	.000	.000	.000	1.000	1.000	1.000	.999
Keno – Spot Ticket – 2.2 to 1	.25	20%	50	.000	.002	.025	.164	.836	.500	.164	.025
			100	.000	.001	.014	.067	.935	.790	.546	.210
			500	.000	.000	.000	.001	.999	.998	.996	.987

that even the most optimistic, forceful, moderately wealthy, or even "lucky" gambler cannot hope to repeal.

There is an additional and somewhat surprising conclusion to be drawn from Table 15. Comparing the corresponding probabilities for the two roulette bets we see that the higher odds (longshot) bet affords larger probabilities of being ahead in every case. Also it is clear that the probability of being ahead on any particular type of bet drops as the number of bets increase (look at the "ahead 0 or more dollars" column). Both of these observations lead us to the following conclusion. If you are fighting unfavorable odds and are interested in winning a specific sum of money while wagering a fixed total amount, your best strategy is to make large bets and, under equal expectation, to prefer higher payoff lower probability types of bets. This will result in fewer bets (trials) and increased probability of achieving your goal. Is there a catch in this strategy? Yes, for a large bet, longshot-biased strategy also increases your probability of losing larger amounts and losing quickly!

Betting systems

Despite a rational interpretation of probabilities, expectations, and even the binomial distribution, there are many gamblers, both casual and inveterate, who believe that they have found or will find a winning *system* of betting in games with a house edge. Most of these proposed systems are naive, based on superstition, humorous, or sometimes pathetic. There are certain systems, however, which are compelling, difficult to deflate, and perhaps even successful for specific gambling goals (lowering the house edge not being one of them). We will not consider systems based upon elaborate record keeping to spot "non-random" trends of the wheel, dice, or whatever. Most such efforts depend on after-the-fact reasoning and are doomed to failure, though cases of faulty (or crooked) apparatus have very occasionally led to success. We also dismiss systems for "knowing" what is due or for recognizing "hot" dice and "friendly" wheels until such time as parapsychology finds itself on a firmer foundation. The systems we do consider base their often considerable appeal on varying the bet size in some fashion depending on what has happened previously. We analyze several such systems knowing full well that there are many other "sure-fire" systems not treated here.

A most intriguing system is the "double when you lose" or *Martingale* strategy. In this system one starts betting at a given stake, say 1 unit, and doubles the previous bet after a loss while returning to the original 1 unit stake after each win. It is easy to see that after r wins the player will be ahead by r units , and that the only thing the player need fear is a long streak of consecutive losses. Indeed if a player goes to Las Vegas with the primary goal of coming out ahead (no matter by how little), it is hard to imagine a better system. If his bankroll is $63 and a $1 minimum bet is in effect, he should play the doubling system starting with a $1 even payoff bet, planning to quit and go home as soon as he wins the first time. The only way he can fail is to lose the first 6 bets $(1 + 2 + 4 + 8 + 16 + 32 = 63)$ which has probability $(1 - p)^6$ (where p is, as usual, the probability of winning on each trial). Even for Las Vegas roulette ($p = .474$) the probability of losing the first 6 bets is $(.526)^6 = .021 (\approx 1/50)$, a comfortably small magnitude. It is true that his loss of $63 in this unlikely event will be much greater than his hoped for $1 gain, but is not the payoff (bragging to all one's friends about how he beat the Vegas syndicate) worth the small risk? The answer is *yes* if winning something (never mind the air fare) is his main goal. The answer is *no* if he believes that his mathematical expectation per dollar bet has been altered. The system would be foolproof but for two vital facts:

1. The player has only a finite amount of capital.
2. The casino imposes a maximum on any given bet.

Each of these facts imposes a limit upon the number of losses beyond which the doubling must be abandoned.

Assuming that the doubling system can only be followed n times and taking $p = .5$ (a fair game), let us apply the expectation concept to this "go home a winner" system. Then

$$p(\text{losing on all } n \text{ bets}) = q^n,$$

and hence

$$p(\text{winning on one of the first } n \text{ bets}) = 1 - q^n.$$

Since the payoff after the first win is 1 unit while the payoff after n losses is $1 + 2 + 4 + 8 \ldots + 2^{n-1} = 2^n - 1$ units, we have

$$X = (1 - q^n) - q^n(2^n - 1) = 1 - q^n 2^n.$$

In the special case $p = .5$, we have $q = 1 - p = .5$ and $X = 0$. Thus, as expected, the doubling strategy does not affect the expectation per dollar bet in this special case.

The class of *cancellation systems* is also fascinating to analyze. Decide upon how much you want to win and then write down a list of positive numbers (the list may be of any size) whose sum equals the amount you

want to win. At any stage, your next bet is the sum of the first and last
numbers currently on the list (if there is just one number left, make that
your bet). If you win, cross the two (or conceivably one) numbers just
used from the list. If you lose, write the amount just lost on that bet at
the end of the list. Continue until all numbers (old and new) are crossed
from the list (in which case you have achieved your goal) or until you
are broke! As an example, let us set out to win $21. In Table 16 we keep
a tally of what transpires based on some plausible but fictitious outcomes. Our initial list, chosen to add to 21, is 4, 7, 1, 3, 4, 2.

TABLE 16

A Cancellation System in Action

Current List	Amount Bet	Outcome	Current Overall Profit
4, 7, 1, 3, 4, 2	6	Lose	− 6
4, 7, 1, 3, 4, 2, 6	10	Lose	− 16
4, 7, 1, 3, 4, 2, 6, 10	14	Win	− 2
7, 1, 3, 4, 2, 6	13	Lose	− 15
7, 1, 3, 4, 3, 6, 13	20	Lose	− 35
7, 1, 3, 4, 2, 6, 13, 20	27	Win	− 8
1, 3, 4, 2, 6, 13	14	Win	6
3, 4, 2, 6	9	Lose	− 3
3, 4, 2, 6, 9	12	Lose	− 15
3, 4, 2, 6, 9, 12	15	Lose	− 30
3, 4, 2, 6, 9, 12, 15	18	Win	− 12
4, 2, 6, 9, 12	16	Win	4
2, 6, 9	11	Lose	− 7
2, 6, 9, 11	13	Win	6
6, 9	15	Lose	− 9
6, 9, 15	21	Win	12
9	9	Lose	3
9, 9	18	Win	㉑ goal achieved
Total	271	8 Wins 10 Losses	

It can be seen why the overall profit equals the sum of the numbers in
the initial list. Every loss simply adds to the list's sum, while each win
removes from the list the amount won. At any stage the sum of the
numbers in the list represents the amount we have yet to win to achieve
our goal. The example and more general analysis show why this system
seems so appealing. As long as losses do not outnumber wins by close to
a 2 to 1 margin, more numbers will be crossed out (2 for each win) than
will be added (1 for each loss), so the list will shrink. As with the

doubling system, the flaw in the system is that bet sizes may escalate, reaching your or the house's limit. Specific mathematical analysis is difficult (Exercise 7 provides a start), but the argument in the next paragraph should shatter any false hopes the system player might have for this or any other variable bet system on a fixed probability game (but see the following section on a genuinely winning system in blackjack).

The expectation calculations of Chapters 2 and 3 have hopefully convinced the reader that any single bet of a given size in a house edge game is not a winning system. It is a small step from this to the realization that any repeated series of *fixed sized* bets cannot alter the house edge. Consider now any variable bet system. Despite its possible numerical mystery and the ordering of the various bets, it is simply made up of groups of these fixed sized bets (a certain number of $1 bets, some $2 bets, etc.). Since each group of fixed sized bets cannot alter the house edge, neither can any combination of such groups. While these variable bet systems are fun to tinker with, we need not trouble ourselves any longer about their possible success. The casinos do not trouble themselves either. They welcome and thrive upon system players in roulette, craps, and any other game where the edge is constantly in the casino's favor.

A brief Blackjack breakthrough

The game of blackjack, like craps, has two versions—a "friendly" version where the rotating dealer is free to make his own decisions and a casino version (sometimes called twenty-one) where the house dealer must follow a simple and inflexible strategy. We confine ourselves to the casino version of blackjack, whose rules we assume the reader knows or is willing to find out by undergoing a quick course with a knowledgeable dealer, a real deck, and monopoly money.

Our interest in blackjack here stems primarily from the fact that there currently do exist playing and betting systems which can make this game a favorable one for the casino player (the only commonly known example). It is important to see why this does not contradict our analysis in the last section. Indeed repeated play at blackjack is *not* a binomial distribution situation since the probabilities change, depending on what cards have already been dealt. While a deal of blackjack starting with a full deck gives the dealer a well-documented house edge (1 or 2 percent), the dealing of cards without replacement allows for situations when the deck is favorable to the bettor. We illustrate with two contrived but instructive situations.

Blackjack example 1. Imagine that by careful observation and prodigious counting you and the dealer start a new hand with just 5 cards remaining in the deck, known by you to be 3 eights and 2 sevens. Your strategy: bet the house limit, stand pat, sit back, and smile. Since the dealer *must* take a card with 16 or under it is easy to see that, no matter which two cards you have, the dealer must take a third card which will put him over with at least 22.

Blackjack example 2. After initial bets have been made and up cards dealt, an *insurance* bet of up to one half the initial bet can be made by a player when the dealer's up card is an Ace. The player is paid 2 to 1 on this bet if the dealer has blackjack, and loses otherwise. It is indeed a form of insurance against a dealer's blackjack. Suppose you have seen 5 cards from the deck which include, besides the dealer's Ace, 4 cards, all nines or below. Then p(dealer has blackjack) $= 16/47 \approx .34$. An insurance bet of 1 unit at 2 to 1 house odds will then lead to

$$X(\text{insurance bet}) = \frac{16}{47}(2) + \frac{31}{47}(-1) = \frac{1}{47} \approx .02 \quad [2\% \text{ player's edge}].$$

Since this expectation is positive, you should take as much insurance as you can in this not so unrealistic situation.

The above examples should illustrate how varying one's bets can be a good strategy in blackjack. It is remarkable that if this strategy is carried through to the limit, blackjack can be made into a game with a significantly negative house edge. It is first necessary to know an effectively optimal strategy for full deck blackjack and then, taking into account cards that are seen, to vary play slightly and bet sizes considerably. Roughly speaking, the richer the deck is in tens and face cards, the better off the player is and the more he should bet. The more 5's and 6's there are in the deck, the less he should bet (can you see why?). It is now blackjack folklore how the mathematician E. O. Thorp used a digital computer to evolve one such winning blackjack system. We heartily recommend Thorp's book *Beat the Dealer* (2nd edition) to the reader interested in his story and his refined and still favorable system.

Needless to say, the prospect of casino gambling with a positive expectation has led to considerable interest and activity involving even more refined systems. The most successful and enterprising approach to date is a hierarchy of five increasingly complex systems developed and marketed by Lawrence Revere. The most refined Revere system assigns a point count to virtually every card (whatever the number of decks) for the purpose of optimizing blackjack card decisions and betting. The cost of the revered and coveted booklet describing this exceedingly difficult system is a mere $200. It is claimed to produce a player edge of 6 percent with only modest variations in bet size.

How did the world of organized gambling respond to the existence of such genuine systems? After initial scepticism (a natural reaction in view of the many poor systems previously tried), the casinos became believers. They responded with minor rule changes, multiple decks, more frequent shuffling, and eviction of winning players suspected of effective card counting. The edge has been nullified for all but the cleverest, quickest, and most subtle card counters. Even allowing for the small, roving band of highly skilled and seldom detected blackjack system players, the casinos as usual have increased their edge. For, much to their delight, they are inundated with pseudo-system players attracted by the gambling urge and the publicity that the genuine systems have generated. Should the supply of skilled system players grow, however, be on the lookout for another rules change—the casinos must have their edge.

Exercises

5.1   An honest die is rolled 5 times.

a) Compute the probability of obtaining (i) 0 sixes, (ii) 1 six, (iii) 2 or more sixes.

b) If you are given $10 whenever 2 or more sixes occur, how much should you pay each time you play this 5 roll game in order for it to be fair?

5.2   Four sets of tennis are played by two evenly matched players. Is the set score more likely to be 2-2 or 3-1 when they finish? Justify your answer mathematically and be sure to discuss what simplifying assumptions you make to deal with the problem mathematically. Are your assumptions realistic?

5.3   Chapter 1 quotes a passage from Dostoyevsky's *The Gambler* in which a roulette player observes that throughout the previous day *zero* came up only once on a particular wheel. Recall that $p(\text{zero}) = 1/37$ on a European wheel and assume for this problem that a wheel is spun exactly 370 times a day.

a) Using the binomial distribution, write an expression for the probability of 1 or fewer zeros in a given day's play of a European roulette wheel. Given that $(36/37)^{369} \approx .00004066$, show that the value of this expression is .000446.

b) Use the normal approximation to the binomial distribution and Table 14 to estimate the probability of obtaining 1 or fewer zeros in a given day's play. Explain why this answer differs from the true value given in a).

5.4   Consider a game with an even money payoff and a probability p of winning each play of the game. Suppose the game is to be played n times. Use the formula for Z to prove that p(ending ahead $10 using $1 bets) $= p$(ending ahead $50 using $5 bets).

5.5   a) Use the normal approximation to the binomial distribution and Table 14 to check each of the following entries in Table 15:

(i) p(ending ahead 0 or more dollars after 100 even money $1 craps bets) $= .48$.

(ii) p(ending behind 20 or more dollars after 500 roulette bets of $1 at 8 to 1) $= .56$.

b) Explain why the entries in each row of Table 15 do not add up to 1. Then explain why the two $D = 0$ entries in each row sometimes add to 1 (as in Keno and 8 to 1 roulette bets), but usually do not.

c) Explain how Table 15 can be used to determine p(ending up exactly even) and obtain this probability for

(i) 100 bets in a fair game and even money.

(ii) 50 bets at 8 to 1 in roulette.

5.6   Suppose the doubling Martingale system described on pp. 78–79 is to be applied on an even money bet having probability p of success and $q = 1 - p$ of failure. Suppose also that you plan to start betting at 1 unit, that you plan to quit after your first win in order to "go home a winner", and that you cannot absorb more than n consecutive losses.

a) Prove that the expectation for the *total amount you will bet is*

$$p + 3pq + 7pq^2 + 15pq^3 + \cdots + (2^n - 1)pq^{n-1} + (2^n - 1)q^n.$$

b) Show that the above amount simplifies to $(1 - 2^n q^n)/(1 - 2q)$. Hint: Substitute $1 - q$ for p, then simplify and sum a geometric progression.

c) Conclude that, even though your probability of "going home a winner" is $1 - q^n$, your expected winnings *per unit bet* (i.e., expected winnings $\div$ expected total bet) are $1 - 2q$, precisely what it is for a single 1 unit bet in this game.

5.7   Consider the simple cancellation system described earlier. Suppose we start with s numbers to be cancelled and a conveniently large amount of cash. Let $w(n)$ and $l(n)$ be the number of wins and losses in the first n trials. (Thus $w(n) + l(n) = n$.)

a) Show that play terminates as soon as

$$w(n) \geqslant \frac{s + l(n)}{2}.$$

b) Show that play will eventually terminate if we assume that

$$\frac{w(n)}{l(n)} \geqslant \frac{1}{2} + E \qquad \text{for all } n,$$

where E is any positive constant. This says in particular that success must result if, as the number of bets gets larger and larger, the ratio w/l has a limit exceeding $1/2$.

5.8   You have seen 12 cards from a 52 card blackjack deck. The cards seen are the dealer's ace now showing, 4 face cards (10, J, Q or K), and 7 other low cards (9 or lower).

a) Is insurance a good bet in this situation? Explain by computing the expectation on a \$1 insurance bet.

b) Suppose you already have a blackjack (with your face card and ace included among the 12 cards you have seen) and make an initial \$2 bet. Compute your expectation on this hand *with* a \$1 insurance bet and then *without* an insurance bet. Comment on your results. Note: If you have blackjack but the dealer does not, you win \$3 for every \$2 wagered. If both you and the dealer have blackjacks, then your wager is returned without profit.

5.9   Assume that when a blackjack deck is "unfavorable" your probability of winning a hand is .48, and that when the deck is "favorable" your probability of winning a hand is .54. Suppose further that you bet \$1 whenever the deck is unfavorable and \$5 when it is favorable. Finally, suppose p(deck favorable) $= 1/4$ and p(deck unfavorable) $= 3/4$. Also ignore increased winning ($3:2$ odds) on blackjacks and ignore the advantages of doubling down and splitting pairs.

a) Compute the mathematical expectation per \$1 wagered (i.e., expected winnings ÷ expected bet) in the above situation, and compute the *house* edge. You will need to show that your expected bet on each hand is \$2.

b) Discuss ways in which the above assumptions are unreasonable. Does the unreasonableness invalidate the point illustrated by a)?

5.10   The Resorts International Casino in Atlantic City allows an added option in blackjack known as *surrender*. A player may surrender after inspecting his first two cards and the dealer's "up" card, in which case he withdraws half of his bet and gives the other half to the house. Let p be the player's probability of winning the hand at the time when he must decide whether or not to surrender. For simplicity, ignore the possibility of ties.

a) Compute for a \$2 bet the player's X(surrender) and X(don't surrender). Do the same for a bet of r dollars.

b) Conclude that, assuming p can be accurately estimated or computed, surrender is a mathematically wise course of action only when $p < 1/4$.

CHAPTER   6

Elementary Game Theory

What is game theory?

In the preceding chapters we have considered a variety of specific games and the specialized reasoning that goes with the analysis of each. Many of the games discussed, including all of the casino games, involve a single player pitted against a randomizing device. These are picturesquely referred to as "games against nature." The subject of game theory sheds light on such games, but derives its main thrust from contests involving two or more self-maximizing players each having a variety of choices and often conflicting interests. Our development will just scratch the surface of the formal "Theory of Games" which sprang almost full blown from the minds of mathematician John von Neumann and economist Oscar Morgenstern. Since the appearance of their book *Theory of Games and Economic Behavior*, game theory has had tremendous impact upon quantitative social science and a most interesting history. Hailed after this book's publication in 1944 as the long awaited conceptual framework which any real (deductive) science needs, the theory was expected to do for the social sciences (primarily economics, political science and psychology) what calculus did for the physical sciences. Game theory study and research was encouraged, embraced, and generously funded; but interacting groups of people do not behave like atoms, molecules, or even billiard balls. Inevitably, after so much initial enthusiasm, disillusionment set in, fuelled by game theory's excessive claims, overquantification and dehumanization of real life situations, and overidentification with military problems. Since the early 1970's a happy equilibrium has been reached, and game theory is enjoying a revival in mathematics and the social sciences.

The abstract entities studied in game theory come in various sizes and forms. In terms of the players the games can be 2-*person* or *n-person*

($n \geq 3$) with payoffs having *zero sum* or *nonzero sum*. For purposes of analysis, games can be described in *extensive form, normal form,* and *characteristic function form*. One of our goals in this chapter will be to define and explain these various types of abstract games. This is done by concentrating primarily on the strategic analysis involved in the 2-person theory. We complete the chapter by touching lightly upon the rich theory of *n*-person games, considering some aspects of how power might be modeled and computed in certain group voting situations.

The reader will soon realize that, despite the beautiful and very general reasoning employed in game theory, the actual examples which can be feasibly considered are a far cry from most "real" games. Indeed, game theory applies to actual game playing in much the same way as probability and expectation theory apply to complex gambling situations (perhaps even less so). The primary value of game theory is that it lays bare the nature of interpersonal cooperation, competition, and conflict, giving treasured insights into the elusive question of what constitutes rational thought and behavior.

Games in extensive form

By a game in this chapter we shall mean the specification of a set of players, rules for playing the game, and an assignment of payoffs or utilities to all possible endings resulting from the various actions of the players. In this section we will be more specific, requiring the rules to specify the order and consequence of *moves* for the various players, how the game ends, and who gets what under the various endings. Games described by such detailed, move-by-move information are said to be in *extensive form*.

Many games, backgammon being an excellent example, also depend upon results of a randomizing device. Such games can be viewed as having an additional player, called *chance*, with its moves and their consequences specified just as for the other players. Analysis of such games requires, in addition to the techniques to be presented in this chapter, repeated probability and expectation calculations of the type considered in earlier chapters. In order to concentrate on new ideas, we will not discuss games with chance moves any further.

Before making further definitions we consider a very simple example:

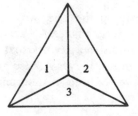

The game of Nonsense: In the figure on p. 87 player A first chooses an interior triangle (1, 2, or 3), and then player B chooses one of the remaining two triangles. Player A must then choose the last triangle. The payoff to each player is simply the sum of the triangle numbers chosen by that player.

It is obvious what *should* happen in this game, but to illustrate where we are headed we analyze all possible ways this game *could* be played out. Our vehicle for this analysis is the *game tree*, which graphically illustrates all possible moves at each stage of the game.

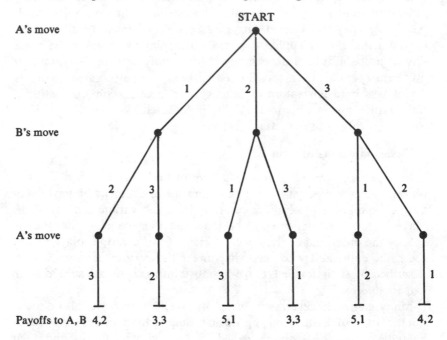

The starting node or *root* of the tree branches downward (the way trees often grow in mathematics) into three edges, indicating the three moves available to A at this stage. For each of these moves B has two options. After A's forced second move, the game ends. We see from the tree that there are 6 possible ways for a game of Nonsense to be played out. The 6 pairs of payoffs are indicated below the extremities of the tree.

By working backwards up the tree, we can find the best move for each player at any given turn. Player A's second turn leaves no choice, but B can see by looking at the payoffs that he should always choose the right branch for his move. Player A, also able to see and reason that B should do this, can use the tree to determine what his best opening move will be—clearly the tree's initial right hand branch corresponding to a choice of triangle 3.

We now introduce the important concept of a *strategy* for a player in a game. Intuitively, a strategy for a player consists of a *complete* specification, in advance, of what he will do at every node of the tree where he has to make a choice. Thus player A has 3 possible strategies in Nonsense (his second move requires no specification). We denote the 3 strategies for player A by A_1, A_2, and A_3, where the subscript in this case denotes the triangle initially chosen. Player B has 8 different strategies, which we list in a shorthand code to be explained below.

B_1:	$A1 \Rightarrow B2$;	$A2 \Rightarrow B1$;	$A3 \Rightarrow B1$
B_2:	$A1 \Rightarrow B2$;	$A2 \Rightarrow B1$;	$A3 \Rightarrow B2$
B_3:	$A1 \Rightarrow B2$;	$A2 \Rightarrow B3$;	$A3 \Rightarrow B1$
B_4:	$A1 \Rightarrow B2$;	$A2 \Rightarrow B3$;	$A3 \Rightarrow B2$
B_5:	$A1 \Rightarrow B3$;	$A2 \Rightarrow B1$;	$A3 \Rightarrow B1$
B_6:	$A1 \rightarrow B3$;	$A2 \Rightarrow B1$;	$A3 \Rightarrow B2$
B_7:	$A1 \Rightarrow B3$;	$A2 \Rightarrow B3$;	$A3 \Rightarrow B1$
B_8:	$A1 \Rightarrow B3$;	$A2 \Rightarrow B3$;	$A3 \Rightarrow B2$

Each strategy has three specifications. Thus $A1 \Rightarrow B2$ tells B's response (triangle 2) to A's initial choice (triangle 1). Strategy B_3, for instance, requires that player B choose the next triangle in a clockwise direction from the one chosen by A. Clearly B_8 is the best or *optimal* strategy for B, but we stress that the term strategy alone does not carry with it any assumptions about whether it is wise or foolish.

We mention an added subtlety which can arise in constructing and interpreting the game tree for certain games in extensive form. If players reveal their moves or decisions simultaneously, this may not be reflected in the top to bottom ordering of a game tree unless added designations are made. Consider the game of matching pennies, in which player A secretly selects "head" or "tail" and player B must guess (match) A's selection to win. If we describe the game tree by the diagram on p. 90 the impression might be created that B could observe A's move, leading to an obvious strategy for B. In general, suppose it is a player's move, and he does not know exactly where he is in the game tree, but only that

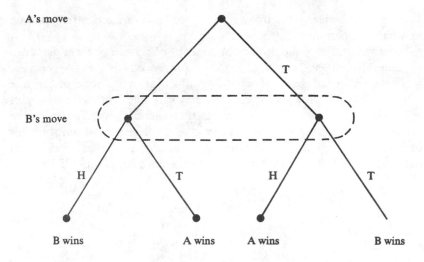

he is at one of the nodes of a set S. Then we lump the nodes of S together by enclosing them with a dotted oval. Such a set is called an *information set*. To extend our notion of a player strategy to this situation, the player must specify, in advance, one move for each of his information sets in the game tree. Games having trees for which each node is a separate information set (so no identification is necessary) are called *games of perfect information*. A bit of thought should make it clear that chess and backgammon are in this category, while poker and bridge are not.

At the risk of dragging out this extensive discussion of game trees, we briefly consider the shape of the tree for the game of Tic Tac Toe. If we ignore symmetry, the game tree starts out as indicated in Figure 5 ("X" is the player going first). It should be apparent from the complexity of the Tic Tac Toe tree that description and treatment of particular games in extensive form is often not feasible. Thus even a game as simple as Tic Tac Toe (intelligent ten year olds usually intuit optimal strategies for "X" and "0" which lead to a forced draw) can be played out in a vast number of ways and with a dizzying number of strategies (see Exercise 6.1).

The number of possible plays and strategies for more "serious" games such as checkers and chess is truly astounding. We shall remark on this further in the chapter's final section, where games trees will be used to illustrate how computers can be programmed to learn from experience. For now we focus on a second and often more workable formulation of games.

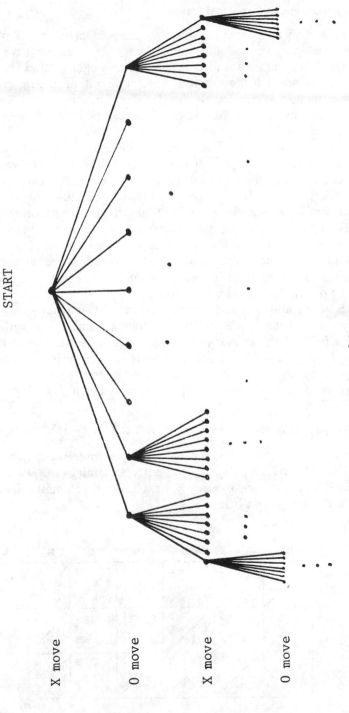

Figure 5   Skeleton of tic tac toe game tree.

START

X move

0 move

X move

0 move

Two person games in normal form

In this section we restrict attention to games involving just 2 players and assume that the possible strategies for each player have been enumerated. It is important to note that such an enumeration is always possible in theory, but the overwhelming proliferation of strategies for even the simplest games (recall the Tic Tac Toe discussion) imposes severe limitations on the class of games that can be handled this way in practice.

In a game with no chance moves, once the two players choose particular strategies, the play is completely determined. We have already assumed that this play results in a certain numerical *utility* or payoff to each player. In most common games this utility might be the value of winning to the winner and the (negative) value of losing to the loser. Here we allow complete freedom as to what these utilities might be (we can have outcomes in which both players gain, both players lose, etc.), but we assume *complete information* for each player about what the two utilities for each outcome are. We avoid some very fundamental questions of how utilities can be defined.

Much of the above should become clearer through some examples. We give one example which derives from a game given earlier in extensive form. Consider the Nonsense game introduced in the previous section. To make the game more interesting we change it by redefining the utilities or payoffs for the various outcomes as follows:

Let x be the number on A's first triangle and y the number on B's triangle. Then

$$\text{pay off to A} = x - \frac{3}{y}, \qquad \text{Payoff to B} = (-1)^{x+y}y$$

i.e. pay off to B is y if $x + y$ is even, $-y$ otherwise. Recall from p. 89 that A has 3 strategies A_1, A_2, A_3, while B has 8 strategies $B_1, B_2, \ldots, B_8$. We construct a rectangular array with 3 rows, 8 columns (called the payoff matrix) as shown below.

Game 1   (Modified Nonsense)

	B_1	B_2	B_3	B_4	B_5	B_6	B_7	B_8
A_1	-2 / $-.5$	-2 / $-.5$	-2 / $-.5$	-2 / $-.5$	3 / 0	3 / 0	3 / 0	3 / 0
A_2	-1 / -1	-1 / -1	-3 / 1	-3 / 1	-1 / -1	-1 / -1	-3 / 1	-3 / 1
A_3	1 / 0	-2 / 1.5	1 / 0	-2 / 1.5	1 / 0	-2 / 1.5	1 / 0	-2 / 1.5

To see how the entries in the payoff matrix for Game 1 were arrived at, we consider for example the $A_2 B_3$ square. Suppose A plays strategy A_2 and B plays strategy B_3. From our earlier description of the strategies, this means A first chooses triangle 2. Strategy B_3 then requires B to choose triangle 3 since A chose triangle 2. Thus, $x = 2$ and $y = 3$, so we see from our payoff rule that A wins $2 - (3/3) = 1$ unit and B wins $(-1)^{2+3}3 = -3$ (i.e., B loses 3). We write A's payoff below, B's payoff above the diagonal line in the square formed by the intersection of row A_2 and column B_3. The other 23 possible payoff pairs may be verified similarly by repeated reference to the description of strategies A_i and B_j listed earlier. This matrix constitutes our description in *normal form* of the game of Modified Nonsense.

Our subsequent examples are games already in normal form; we ignore their origins (if any) in terms of game rules or extensive form. We give one more preliminary example before beginning the general analysis of games in normal form. Consider Game 2 as defined below.

Game 2

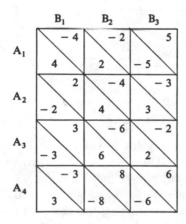

Here A has 4 strategies and B has 3. A typical play might result from A selecting strategy A_2 and B selecting strategy B_3, in which case A wins 3 units and B loses 3 units. The obvious special nature of this example will be the object for analysis in the next section.

Zero-sum games

The two-person situation described by Game 2 has the special but frequently encountered property that for each pair of opposing strategies the sum of the payoffs to the players is zero. Such games are called, naturally enough, *zero-sum games*. They constitute an important special case of the general two-person theory. In this zero-sum situation it is unnecessary to write the payoffs for both players. We adopt the standard convention in the zero-sum case of only writing the payoffs to the row player (in our case A). We remind the reader that the payoffs to the column player are then the negatives of these numbers, and that forgetting this (which is easy to do) will result in incorrect analysis.

Employing this convention, Game 2 becomes

Game 2

	B_1	B_2	B_3
A_1	4	2	-5
A_2	-2	4	3
A_3	-3	6	2
A_4	3	-8	-6

From this matrix we see directly that strategy A_4 is *never* as good as strategy A_1 (no matter what B does). Since A is assumed to be a rational player, he may as well delete A_4 from his list of strategies. Thus the game reduces to

	B_1	B_2	B_3
A_1	4	2	-5
A_2	-2	4	3
A_3	-3	6	2

Now player B, being every bit as rational as A, also sees that A will never play strategy A_4. Thus the juicy 8 payoff of A_4B_2 is no longer available to him. From the reduced version of Game 2 it is now apparent that B_2 is inferior to B_3, so we delete B_2 from the set of B's viable strategy options. Continuing this reasoning for A it can be seen that after B_2 is deleted, A_3 is now inferior to A_2. The reader should check this, as we will henceforth usually jump directly to the final reduced form. Thus we are left with

Reduced game 2

	B_1	B_3
A_1	4	-5
A_2	-2	3

The reader should check that no further reduction is possible.

For one more example, consider

Game 3

	B₁	B₂
A₁	8	− 4
A₂	4	1

Here A examines his strategies and sees that neither appears consistently more rewarding than the other. On the other hand, B sees that B_2 is always preferable to B_1 (recall the need to negate the payoffs). Accordingly the game reduces to

	B₂
A₁	− 4
A₂	①

Being rational, A makes the same deductions as B and therefore always prefers strategy A_2. The conclusion, then, is that rational play will lead to the $A_2 B_2$ strategy pair, with A gaining and B losing 1 unit. By deviating from the B_2 strategy, B will be worse off. Though A could play A_1 in the hope of B's foolishly or accidently playing B_1, this would not be rational or prudent.

We formalize the reasoning in the last several paragraphs as follows. For any player C (either A or B, but remember to negate payoffs for B) with strategies $C_1, C_2, \ldots, C_m$,

C_i is *dominated by* C_j if

> *each payoff* in C_i ⩽ the corresponding payoff in C_j

and

> *some payoff* in C_i < the corresponding payoff in C_j.

Our analysis so far has consisted of deleting all dominated strategies from the game for either player, repeating the search for dominated strategies after each deletion. The end result of this process leaves the game in *reduced form*. If each player in the reduced form has only a single strategy or a set of identical strategies, we say the game has a *pure strategy solution*. Each player is then said to have an *optimal pure strategy*.

There is another procedure for zero sum games which, if successful, leads to an even more dramatic simplification. Again we start with an example.

Game 4

	B_1	B_2	B_3
A_1	-5	11	-7
A_2	-2	8	-1
A_3	-3	-4	14

It should be noted first that there are no dominated strategies for either player. Before abandoning hope for simplification, we engage in the following somewhat elusive but highly compelling reasoning. Player A, reasoning defensively, considers the worst that might happen. A's worst or *minimum* payoffs are -7 if A_1 is played, -2 if A_2 is played, and -4 if A_3 is played. By computing the *maximum* of these minima (-2) and choosing a strategy which yields this maximum (A_2), player A can guarantee a payoff at least as good as this maximum. Naturally B can reason similarly. *Negating payoffs first*, we see that 2 is the maximum of B's minima ($2, -11, -14$), so B's defensive strategy is B_1. It is of great significance that the pair $A_2 B_1$ gives precisely the defensive results to which both players were led. Because of this the strategy pair $A_2 B_1$ can be seen to have high stability in that, if either player deviates from it and the other does not, the deviating player will suffer. Furthermore, if each player deviates from his respective strategy in $A_2 B_1$, then one of the players must gain by returning to it. The reader should check that this stability results because

MAXimum (A's MINima) = $-$MAXimum (B's MINima).

If a game has the property that A's maximin = $-$B's maximin, then any pair of these *maximin strategies* is called a *saddle point* for the game. It follows that a payoff will correspond to a saddle point precisely when this payoff is simultaneously a minimum value in its row and a maximum value in its column. Game 4 has the unique saddle point $A_2 B_1$. If one is able to spot a payoff value in the normal form with this property, the more elaborate maximin procedure will be unnecessary. The maximin procedure is the only reliable way, however, to ensure that a saddle point is not overlooked. Generally a game can have more than one saddle point, but it is easy to check that all saddle points will give rise to the same payoff.

Before leaping to unfounded conclusions, let us return to Game 2 (analysis with Reduced Game 2 will yield equivalent conclusions). In Game 2 we have:

$$A\text{'s maximin} = MAX\{-5, -2, -3, -8\} = -2 \text{ by strategy } A_2$$

$$B\text{'s maximin} = MAX\{-4, -6, -3\} = -3 \text{ by strategy } B_3.$$

Note that these maximin values do not sum to 0 and hence cannot be jointly achieved by any strategy pair. The stability argument that worked so well for Game 4 fails. There is no saddle point here. We summarize:

Maximin principle for pure strategies: In any two-person zero-sum game, if A's maximin $= -$B's maximin, then any strategy pair which gives rise to this joint payoff is a saddle point or solution. Any such solution pair can be used by rational players as pure strategies for the game.

If a zero-sum two-person game has a rational *pure strategy* solution, the maximin procedure will find it. In considering a zero-sum two-person game for analysis we should

1. try to find a saddle point by inspection or by applying the maximin procedure. If a saddle point exists we are done. Otherwise,
2. put the game in reduced form by repeated deletion of dominated strategies. Then,
3. try to find a *mixed strategy* solution.

We now discuss the meaning of step 3 and describe a procedure for implementing this step in the special case where the reduced game has only 2 strategies for each player. Consider Reduced Game 2 which would arise from Game 2 after failure of step 1 and reduction by step 2. We will show that each player should, in successive plays of Reduced Game 2, follow an overall strategy of varying or mixing the given strategies for the game. It is clear that if strategies from game to game are to be mixed, this should be done in some "random" fashion, since otherwise the rational opponent could notice a pattern and exploit it. Let

$$p = A\text{'s probability of playing strategy } A_1$$

and

$$q = \text{B's probability of playing strategy B}_1.$$

Then the following pair of tables gives the probabilities and corresponding payoffs for each of the 4 possible outcomes of Reduced Game 2.

	B_1 (q)	B_3 (1 − q)		B_1	B_3
$A_1(p)$	pq	$p(1-q)$	A_1	4	− 5
$A_2(1-p)$	$(1-p)q$	$(1-p)(1-q)$	A_2	− 2	3

It follows directly from the expectation definition of Chapter 2 that

$$X(\text{player A}) = pq(4) + p(1-q)(-5) + (1-p)q(-2)$$

$$+ (1-p)(1-q)(3).$$

We now perform some algebra on this expectation to put it in a very useful form.

$$X(\text{player A}) = 4pq - 5(1-q) - 2(1-p)q + 3(1-p)(1-q)$$

$$= 14pq - 8p - 5q + 3 \qquad [\text{Collecting like terms}]$$

$$= p(14q - 8) - 5q + 3$$

$$= 14p(q - 4/7) - 5q + 3$$
$$\qquad [\text{Factoring 14 from } 14q - 8]$$

$$= 14p(q - 4/7) - 5(q - 4/7) + 1/7$$
$$\qquad [\text{Forcing a } (q - 4/7) \text{ term and adjusting}]$$

$$= (q - 4/7)(14p - 5) + 1/7 \quad [\text{Factoring } (q - 4/7)]$$

$$X(\text{A}) = 14(q - 4/7)(p - 5/14) + 1/7$$
$$\qquad [\text{Factoring 14 from } 14p - 5].$$

The reasoning now goes as follows. Player A can guarantee himself a 1/7 payoff by letting $p = 5/14$ (he has no control over q, but this "zeros" q out). What if he is tempted to deviate from 5/14, say by letting p exceed 5/14? Then player B, a rational fellow, can discover over a series of games that $p > 5/14$ or $p - 5/14 > 0$. Using the equation for X(player A), B can now make $X(A) < 1/7$. His best approach would be to set $q = 0$ (i.e. always play B_3), in which case the *negative* term $14(-4/7)(p - 5/14)$ would lower A's expectation below $1/7$. Thus A would be unwise to let $p > 5/14$ and, by an obvious analogous argument, would also be unwise to let $p < 5/14$. In summary, A can *do no better* than to vary strategies randomly with $p = 5/14$. Similarly, B should use $q = 4/7$ to prevent A from making $X(A) > 1/7$. (The whole argument is somewhat reminiscent of our maximin analysis). The expected payoff for these mixed strategies is $1/7$ for A and $-1/7$ for B. A careful study of this example should serve as a model for the mixed strategy procedure (step 3) in the reduced 2 by 2 strategy case. Exercise 6.6 considers the general 2 by 2 situation.

We conclude this section by stating without proof the deep and beautiful theorem that every 2-person zero-sum game in normal form has a solution in either pure or mixed strategies. We have seen how such solutions can be obtained through saddle points or (when each person has only 2 strategies) mixed strategy probability calculations. Computation of mixed strategy solutions when players have more than two strategies requires more sophisticated mathematical techniques from a branch of mathematics called linear programming.

Nonzero-sum games and the prisoners' dilemma

A game for which the payoffs to the various players do not always sum to zero is called a *nonzero-sum game*. In this section we briefly consider the two-person nonzero-sum game in normal form. Such games have a rich mathematical theory, but are also subject to considerable philosophical discussion and varied interpretation. After looking quickly at one solution concept for certain of these games, we turn to our main objective, an exposition of the famous Prisoners' Dilemma.

Recall Game 1, which arose from our Game of Nonsense. If we employ the common sense deletion of dominated strategy procedure for player B, it is seen that strategy B_5 is the lone viable strategy remaining for B, giving the reduced matrix

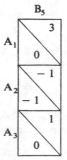

Player A is then indifferent between A_1 and A_3 as far as his best payoff (0) is concerned. What might he do? If he is altruistic he may choose A_1, allowing B to gain 3 units. If he has a sadistic nature he may choose A_3, holding B's payoff to 1 unit at no gain to himself. A little thought suggests another very real option. Player A might politely suggest to B that he is prepared to play A_1, but only if B makes it worth his while with a *side payment* of 1 unit. Such gentle blackmail may or may not pay off, but it does show the added complexity and ambiguity of the nonzero-sum situation. One can be led very quickly into considering concepts such as negotiable versus nonnegotiable games, games with and without side payments, bluffing, etc.

Returning to the original Game 1, we can even throw the cherished idea of deleting dominated strategies into question. Indeed, A_1 is dominated by A_3, so deletion would suggest that only A_3 could ever be of interest to A. This would result in a pure A_3B_5 strategy outcome for the game, not even allowing A's added bargaining strength (blackmail) considered above to surface. The moral we propose is that care and skepticism must be exercised in presenting and analyzing "rational" arguments about nonzero-sum games. We bring this point home with the following hypothetical but, as we shall see, not uncommon game situation.

Two alleged burglars (A and B) are spotted and apprehended running away from the Sparkle Jewelry Store after its burglar alarm has been tripped. Jewelry is found scattered around the store. A search of the suspects turns up no jewelry, but each is found to be carrying a gun. The suspects are then imprisoned in separate interrogation rooms and each is urged to confess to the attempted burglary. Being rational, knowledgeable, and like-minded partners in crime, each evaluates the situation as follows:

a) If both confess they will each get moderate prison sentences (judged by each as a payoff of -4 units).
b) If both refuse to confess they each will get light sentences for carrying concealed weapons—no burglary charge would stick (judged by each as a payoff of 1 unit).
c) If one confesses and the other does not, the confessor, having turned state's evidence, will be set free; the silent one will get a heavy prison sentence, taking the rap for the burglary (judged by each as payoffs of 10 to the confessor and -6 to the "sucker").

They realize then that they each have two strategies (1 = confess, 2 = don't confess) in the following nonzero-sum game.

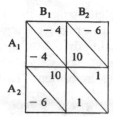

How should the prisoners deal with their dilemma?

If a deletion of dominated strategies is employed, both players arrive at their confess strategies (resulting in a payoff of -4 to each). This strategy pair (A_1B_1) has the compelling maximin stability discussed earlier in that if either player deviates from it on his own, he will suffer. Also, if the players are at any other strategy pair (even (A_2B_2)), at least one player can improve his payoff by *unilaterally* switching to strategy 1. All indications seem to point to A_1B_1 as the inevitable and "rational" end result. But look again! If the players could only cooperate or read each other's minds, would they not readily uphold the "honor among thieves", settling on A_2B_2 and avoiding the unpleasant A_1B_1 outcome? Possibly so, but a little thought suggests that this prospect might be short-lived as each player eyes the possibility of complete freedom provided by being the sole confessor. Each player, suspecting the other, thinking of himself, and being unable to resist the lure of his dominating confession strategy, would lead the pair right back to where they did not especially want to be.

The above is one version of the *Prisoners' Dilemma*, which no amount of philosophical or game-theoretic analysis has yet resolved to the satisfaction of all. Before examining it a bit further we comment that it can be used as a striking though rather oversimplified model of the arms race. Here the players are a pair of military superpowers (with no "third world" competition). Each must decide yearly whether to curtail armament spending or to increase it. If both powers increase (A_1B_1) they have gained relatively little militarily and are poorer financially. If both curtail spending (A_2B_2) then they maintain their arms balance and can use the money saved for more worthwhile purposes, leaving themselves (and the rest of the world) better off. In the remaining two cases the power increasing armaments gains military superiority and a large positive payoff despite the cost, while the "dove" power loses the arms race and a sizable payoff in prestige and power balance. Similar analysis can be given to model or explain price wars, ecologically unsound societal behavior, and other "selfish" action. We thus have a game-theoretic explanation of why irrational group behavior can be expected in certain situations.

We now consider what might happen in *repeated play* of a prisoners' dilemma type game (just as the arms spending game gets replayed every year). It would seem reasonable that the players, looking ahead to their *overall payoff* over the series of replays, would now be able to come to some mutually advantageous agreement. To be specific, imagine that the players in a prisoners' dilemma game know in advance that the game will be played precisely 15 times. Then perhaps at the outset (if preliminary discussion were allowed) or after a few games (if it were not) the players would see that it was in their individual and mutual long-term best interest to cooperate (i.e., play strategy 2). Being rational though, each would think ahead to the 15th and final game. In this game, the question of *future* overall payoffs is no longer an issue, so players will once again revert to their dominant strategy behavior. Having no control over what the opponent will do, each player is better off playing strategy 1; so cooperation on game 15 cannot be expected. What about the 14th game? With the 15th game "up for grabs" this 14th game is now effectively the last game in which cooperation might be reasonable. Again each player reasons that there is little point in unilaterally playing strategy 2. Of course the regression continues until even the first game of any finite series provides no incentive for cooperation. We leave to the reader the pleasure of contemplating what might happen in a game with potential for infinite replay and how a player might increase overall payoffs by "playing dumb" or "accidentally" lapsing into strategy 1 upon occasion.

Simple *n*-Person Games

Our approach to game theory so far has concentrated on games involving only two players. If more than two players are involved in a game, several new difficulties become apparent. Not only does the already substantial complexity of nontrivial games in extensive and normal form increase considerably, but the question of whether and how players may cooperate and band together in coalitions becomes a major issue. Our earlier response to the complexity problem was to bypass the intricacies of the tree structure of a game by directly considering strategies and the normal form of a game. We carry this one step further now by replacing the enumeration of individual strategies by information as to what the various subsets of players can gain by forming *coalitions*. These are the so-called *cooperative games* in *characteristic function form*. We consider the even more special case of *simple games* where only two payoffs are possible, called *winning* and *losing*; and it must be specified in advance precisely which coalitions can achieve the winning payoff.

The most familiar game of the type described above is the n-person *majority game* M_n, where more than half the players (eligible voters at a meeting) must vote in favor of something before it can win (pass). There are two useful ways to describe this game M_n, and we shall need them both. Consider for example the 5-person majority game M_5. It can be represented as the *weighted voting game* $[3; 1, 1, 1, 1, 1]$, where, in general, $[q; w_1, w_2, \ldots, w_n]$ is defined as follows: The subscript n is the number of players. The number q is the *quota* for the game, while the numbers $w_1, w_2, \ldots, w_n$ are the *weights* of the players. By definition, a set S of players is a *winning coalition* of $[q; w_1, w_2, \ldots, w_n]$ if and only if the sum of the weights is at least q. Clearly the winning coalitions of M_5 (where each player has weight 1) are just the subsets of players with 3 or more members. Thus, the *simple game* representation of M_5, obtained by listing all its winning coalitions, is

$$\{ ABC, ABD, \ldots, CDE, ABCD, \ldots, BCDE, ABCDE \},$$

where we have denoted the five players by A, B, C, D, E. We leave it to the reader to check, using the combination ideas of Chapter 4, that there are $C_{5,3} + C_{5,4} + C_{5,5} = 16$ winning coalitions for this game.

We give several more examples of games in weighted voting and simple (winning coalition) forms. Imagine a situation with one "big" player A and two "little" players C and D, where player A requires the help of either C or D (or both) to achieve "success." Players C and D cannot succeed on their own, but they do have some clout in the sense that they can join to prevent A from achieving success. The resulting game BL^2 can be defined as follows:

Weighted voting form: $BL^2 = [3; 2, 1, 1]$

Simple form: $BL^2 = \{ AC, AD, ACD \}$.

The United Nations Security Council has at any given time the five "big powers" and ten other nations as its voting membership. Passage of a resolution (assuming no abstentions) requires support of *all five* big powers and at least four of the others. There are many ways to choose weights and a quota to represent this situation and we give one of them.

Weighted Voting Game:

$$UNSC = [39; 7, 7, 7, 7, 7, 1, 1, 1, 1, 1, 1, 1, 1, 1, 1].$$

It should be clear that the full listing of winning coalitions of UNSC in simple form is exceedingly tedious. Indeed, the number of minimal (9 member) winning coalitions is already $C_{10,4} = 1260$.

The preceding example shows the economy of representing a simple game in a weighted voting formulation when possible. The following example gives a collection of winning coalitions whose game cannot be described by means of quotas and weights (see Exercise 6.8). The winning coalitions appear in dictionary order.

Simple Game:

$$SG = \{AB, ABC, ABD, ABE, ABCD, ABCE, ABDE, ABCDE,$$

$$ACD, ACDE, BCDE, CDE\}.$$

We shall always assume that any coalition containing a winning coalition is also winning. In this case, simple games can be described by listing only the *minimal winning coalitions* (MWC's). These are coalitions which are winning, but contain no smaller winning coalition. The MWC's for the above simple game have been set in bold type. The minimal winning coalitions will play an important role in our forthcoming definition for the power of a player in a simple game.

Power Indices

Our goal for the two person games considered earlier was to determine what moves and strategies players should pursue under assumptions of rational thought and behavior. Even for the simple class of n-person games we are considering, such a task is often not well-defined and almost always beyond our capabilities. Instead, game theorists seek to determine a *value* for the n-person game representing how rational players might apportion the total payoff available in the game. For a simple game this value can be interpreted as the *percentage of power* each person in the game might possess. Accordingly, we define a *power index* on the class of simple n-person games as a procedure which assigns, for each such game, a number to each player representing the player's power in the game. Since the numbers are percentages, we require that they sum to one. In symbols, each simple n-person game G is assigned a power index $p(G) = (p_1, p_2, \ldots, p_n)$ where $p_i =$ the power of player i, and $p_1 + p_2 + \cdots + p_n = 1$.

Power indices are not only an interesting theoretical idea, but they have been used in law courts to settle disputes involving fair representation in voting bodies. There are several well established power indices in game theory, most notably the original index defined by game theorist Lloyd Shapley and economist Martin Shubik and a later index developed by the lawyer John Banzhaf. In our brief development we use a recent power index developed by the author in collaboration with political scientist John Deegan, Jr. The decision to use this index is based not only on personal bias, but also on the fact that computation of the index values is more straightforward and illustrative of some of the mathematics developed in earlier chapters.

Consider the game BL^2 with players A, C, and D as introduced in the preceding section. It is easy to check that there are precisely two MWC's in BL^2, namely AC and AD. Let us assume that *only* MWC's are allowed to form, that they form with equal probability (in this case each forms with probability $1/2$), and that players belonging to the MWC which forms divide the 1 unit payoff equally. Then the expectation for each player can be found by what are now, hopefully, familiar procedures. It is this expectation which we use to define the power of a player. For the game BL^2 we obtain the following results:

$$p_A = X(\text{player A})$$

$$= p(\text{AC forming})(\text{A's share of AC})$$

$$+ p(\text{AD forming})(\text{A's share of AD})$$

$$= \frac{1}{2} \cdot \frac{1}{2} + \frac{1}{2} \cdot \frac{1}{2} = \frac{1}{2}.$$

Similarly,

$$p_C = X(\text{player C}) = p(\text{AC forming})(\text{C's share of AC}) = \frac{1}{2} \cdot \frac{1}{2} = \frac{1}{4}$$

$$p_D = X(\text{player D}) = p(\text{AD forming})(\text{D's share of AD}) = \frac{1}{2} \cdot \frac{1}{2} = \frac{1}{4}.$$

Thus our power index applied to BL^2 gives $p(BL^2) = (\frac{1}{2}, \frac{1}{4}, \frac{1}{4})$.

Before considering more complex examples, we break the power index computation into three steps. To obtain the power p_i of a player i in a simple game G:

(1) Compute the number g of MWC's and list or describe them.
(2) For each MWC containing player i, compute the *reciprocal* of the number of players in it.
(3) Add all the reciprocals obtained in step (2) and divide the total by g—this gives the power p_i of player i in the game G.

Applying the above method to the simple game SG on page 104, we recall that the MWC's are {AB, ACD, CDE}, so g = 3. Thus we obtain

$$p_A = \frac{\frac{1}{2} + \frac{1}{3}}{3} = \frac{5}{18} \qquad p_B = \frac{\frac{1}{2}}{3} = \frac{1}{6} \qquad p_C = \frac{\frac{1}{3} + \frac{1}{3}}{3} = \frac{2}{9}$$

$$p_D = \frac{\frac{1}{3} + \frac{1}{3}}{3} = \frac{2}{9} \qquad p_E = \frac{\frac{1}{3}}{3} = \frac{1}{9}.$$

Thus $p(SG) = (5/18, 1/6, 2/9, 2/9, 1/9)$; it should be noted that the individual powers sum to unity.

As our last example we consider a simplified version of a political game familiar to most American citizens, the voting procedure for enactment of laws by the United States Congress. The Congress has 435 representatives and essentially 101 "senators" (100 real senators plus the Vice President, who may cast a deciding vote whenever the senate is deadlocked at 50 to 50). Passage of a new law requires "yes" votes from 218 representatives, 51 senators, *and* the President. (We ignore the possibility of a congressional veto override here.) Using the absolute majority rule terminology of the last sections, we see that the congressional game, which we call USA, requires simultaneous winning coalitions in M_{435} (the house), M_{101} (the senate), and M_1 (the President's approval).

In preparation for computing the power of the players in USA, we note that all MWC's are made up of 218 representatives, 51 "senators," and the President. Accordingly, the total number of MWC's in USA is obtained using our notation for combinations as $g = C_{435,218} \cdot C_{101,51}$. Every representative r belongs to $C_{434,217}$ house MWC's and to $C_{434,217} \cdot C_{101,51}$ USA MWC's, each of which has $218 + 51 + 1 = 270$ members. Similarly, each "senator" s belongs to $C_{435,218} \cdot C_{100,50}$ USA MCW's of size 270. The President belongs to *all* of the g different MWC's. Putting all of this together, we obtain:

$$p_{Pres} = \frac{g \cdot \frac{1}{270}}{g} = \frac{1}{270} \approx .00370$$

$$p_s = \frac{C_{435,218} \cdot C_{100,50} \cdot \frac{1}{270}}{g} = \frac{C_{100,50} \cdot \frac{1}{270}}{C_{100,51}} = \frac{51}{101} \cdot \frac{1}{270} \approx .00187$$

$$p_r = \frac{C_{434,217} \cdot C_{101,51} \cdot \frac{1}{270}}{g} = \frac{C_{434,217} \cdot \frac{1}{270}}{C_{435,218}} = \frac{218}{435} \cdot \frac{1}{270} \approx .00186.$$

Since the presidential power value seems much too low, it is appropriate to question our model of power. If we realize that we are measuring only the power to enact legislation, independent of all the other kinds of actual and psychological power a President has, the results may not be totally unreasonable. Other power indices with different assumptions do give greater power to the President and to senators, and it seems likely that the index we present is not appropriate for the U.S. Congress. Nevertheless, it is the ideas and their implementation rather than political realities that we have worked for here.

Games computers play

Earlier sections of this chapter indicated that the application of abstract game-theoretic analysis to all but the simplest of games is likely to leave us hopelessly stuck in a tree or drowning in strategy enumeration. Remarkably, another far more pervasive scientific breakthrough has paralleled the birth and development of game theory, the phenomenon of the digital computer. It is interesting to note that von Neumann made fundamental contributions to both of these seemingly unrelated fields. Would it not be reasonable to hope that computers might provide the means for dealing with the computational complexity called for in the theoretical analysis of games? We briefly consider some general and specific responses to this question.

The subject of *artificial intelligence* deals with how and to what extent computers and other machines can be programmed or constructed to exhibit "thinking-like" behavior. A major subdivision of the burgeoning discipline of computer science, artificial intelligence (or AI) relates to human intelligence in somewhat the way that game theory relates to real world games. We do not expect to construct or program machines to think in the same way that human beings do; but by seeing and thinking about what machines might be made to do, we can raise and sometimes answer important questions about the nature of thought and intelligence.

Our goal in this section is much more modest. Using the game tree ideas of the early parts of this chapter, we shall completely analyze a fairly simple game called HEXPAWN. Not only will the analysis reveal optimal strategies for HEXPAWN, but we shall get a glimpse of how computers can use the game tree to learn from experience.

HEXPAWN is played on a 3 by 3 board, starting with 3 White pawns and 3 Black pawns as pictured in Figure 6.

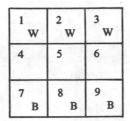

Figure 6   Starting positions for HEXPAWN.

The rules for moving and winning are listed below (we assume W goes first):

• Players move alternately, W downward and B upward.
• A player may move 1 space forward if the space is vacant.
• A player may move 1 space diagonally forward if the space is occupied by an opponent's pawn. Opponent's pawn is then removed.
• These are the only legal moves.

A player wins if any of the following conditions hold:

• One of the player's pawns reaches the opponent's end of the board.
• The opponent is unable to move. (Thus we are not allowing "stalemate".)

At this point we urge the reader to take a brief time out to run through some practice HEXPAWN games. This will provide facility in understanding the following tree construction. It should also start the reader asking that often most crucial of all game questions: "Would I rather go first or second?"

Before drawing the tree, we formalize a rather intuitive notion of strategic analysis. Suppose it is player W's turn to move at a given node in a tree and *all* branches lead to a node (position) from which opponent B can force a win. Then, assuming perfect play, we would say that B controls or captures the original node. On the other hand, if *some* branch from a given node (at W's turn) leads to a node already captured by W, then W should capture the given node. These common-sense observations are schematized in Figure 7.

The first 5 levels of the HEXPAWN game tree are pictured in Figure 8. To prune the tree down to size without losing crucial information, we only list one branch of any symmetric situation (thus there are two opening moves indicated instead of three). The labels on the branches indicate the number on the square moved *from* followed by the number moved *to*. For symmetrical situations only the least numbered branch is drawn. The dotted lines beneath the tree indicate that additional parts of the tree have been omitted. The reader is urged (Exercise 6.11) to

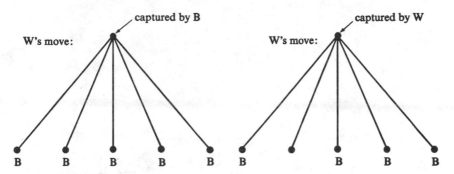

Figure 7 Capturing higher level nodes.

check that the symbol (W or B) to the right of each lower level node correctly indicates the player capturing the node (in accordance with the rules of Figure 7). It is now easy to complete the upward scan of the game tree to show that Black (i.e., the player who goes second) captures the root node, and that Black can therefore *always* win with careful (perfect) play. Indeed, the correct play for Black is simply to choose successive branches down the tree which always join nodes captured by Black. Perfect play for White in this unfortunate case is to hope for Black to make an error and then to stick with nodes captured by White.

The ideas described above can be applied in principle to any game of perfect information in extensive form whose termination rules generate a tree with finitely many nodes and payoffs of win, lose, or draw. After extending the capture rules of Figure 7 to the case of draws (Exercise 6.12), we can work our way up the game tree to discover:

a)   the optimal moves at each node
b)   the player (if any) capturing each node
c)   the player (if any) who can win the game with optimal play.

Applying this tree analysis to chess, for instance, we would be able to discover which of the following statements is true (our "capture" technique implies that exactly one of them must be true):

1.   The White (first) player wins if he plays optimally.
2.   The Black (second) player wins if he plays optimally.
3.   The game always ends in a draw under optimal play.

As a bonus, the game tree tells us precisely what moves are best at each stage of the game—it gives us optimal strategies. To implement this seemingly profound yet obvious result in practice, use of a digital computer is suggested. This brings us back to the subject of artificial intelligence.

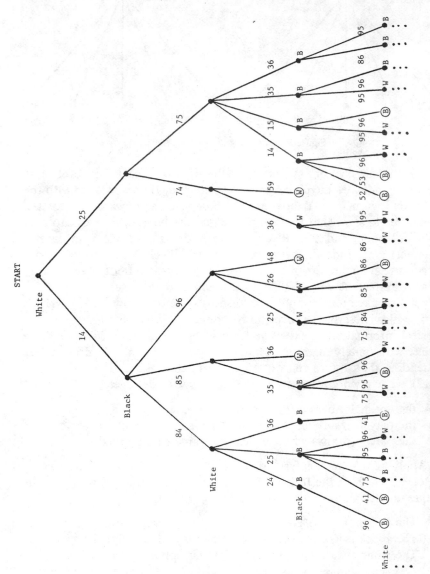

Figure 8   Part of the HEXPAWN game tree.

We are now prepared to strip away the mystery from the idea that a computer can learn through experience. It is fairly easy to program a computer to store and build trees of considerable size and complexity. In programming a computer to play HEXPAWN legally but randomly (a reasonably straightforward task) and to build its game tree as it goes along, the following idea suggests itself. When the computer loses a game it can mark the last branch it took as a "bad" move. (It could also flag branches which led to wins as "good" moves, but this is not necessary.) After losing sufficiently often the computer will be able to avoid all bad moves, an ultimate goal of learning. If the computer then goes second in HEXPAWN it will always play perfectly, having learned all there is to learn. If the computer goes first at this point it will also have learned—namely that it has only bad moves unless its opponent makes a mistake. In summary, the computer can learn to play as well as possible, and it learns most effectively by losing.

If this can be done so easily for HEXPAWN, why not for checkers or chess? The Tic Tac Toe example suggests the problem for nontrivial games—their trees grow so large so fast that even the most powerful computer cannot hope to pursue the tree analysis approach successfully. Indeed, it has been estimated that to construct the game tree for chess to 25 levels (corresponding to looking ahead 25 moves—a rather short game of chess) would require a number of comparisons on the order of 10^{75} (one followed by 75 zeros). Even with a very fast hypothetical access and comparison time (say a billionth of a second), a 25 level analysis would require perhaps 10^{65} seconds of computer time. But the age of the solar system is on the order of 10^{18} seconds! So much for a full tree analysis of chess.

Despite this complexity, increasingly successful computer programs have been written for playing checkers, chess, and other unsolved games. The methods used combine the speed and unerring patience of the computer with board position "evaluation functions" and programmable strategic principles (heuristics). Computers can now play checkers as well as the best human players. Increasingly sophisticated chess programs are raising the level of computer play, though somewhat more slowly than was predicted 10 years ago. The best chess programs can now play on a level with highly skilled amateurs, with inroads being made against lower level experts. One intriguing approach for testing the sophistication of a complex chess program, scarcely possible with human players, is to have the computer play games against itself.

Do these intricate and highly specialized programs serve any purpose beyond the challenge and excitement they provide for the chess enthusiast who competes against or writes them? If we view chess (as we have viewed game theory) as a model for how the human mind might operate at its most rational levels, the rewards of an "optimal" or even expert level chess program might be considerable, yielding deep insights into the nature of human intelligence.

Exercises

6.1   The game of Direction/Color is played as follows. First, player A chooses a direction (Left or Right) and announces it to player B. Then player B chooses a color (Red, White, or Blue). The colors Red, White, and Blue are assigned point values of 2, 3, and 4 respectively. If A chooses Left then A *pays* B the number of dollars then determined by the point value of B's color. If A chooses Right, then B *pays* A the corresponding amount.

a) Draw the full game tree including payoffs for Direction/Color.

b) Explain why player A has only two strategies and fully describe player B's optimal or best strategy.

c) Show that player B has 9 different strategies for the game. [Hint: a single strategy for B consists of specifying one of 3 color responses for each of A's 2 direction choices.]

6.2   Consider a Tic Tac Toe analysis which ignores symmetry and assumes games are continued until all nine boxes are filled.

a) Show that the number of different games that can be played is $9! = 362{,}880$.

b) Argue that there are $9 \cdot 7^8$ different strategies in Tic Tac Toe for the first two moves of the "X" (first) player. [Hint: note that for any opening move there are 8 responses by the "0" player. Now see Exercise 6.1(c) and its hint.]

c) Prove that under our assumptions there are $9 \cdot 7^8 \cdot 5^6 \cdot 3^4 \cdot 1^2 \approx 6.566 \cdot 10^{13}$ different strategies for the "X" player.

d) Write an expression for the number of different strategies for the "0" player.

6.3

	B_1	B_2	B_3
A_1	-4	6	-6
A_2	-1	4	3
A_3	-2	-3	-1

Consider the normal form zero-sum game above.

a) Argue by eliminating dominated strategies that there is a pure pair of strategies (one for each player) for this game. Explain what should happen with rational play, and what payoff will result?

b) Now change the $A_3 B_3$ payoff from -1 to 4. Show that there are no dominated strategies. Then use maximin arguments to show that a pure strategy pair still results.

6.4

	B_1	B_2	B_3	B_4
A_1	2	-4	3	2
A_2	-7	3	3	0
A_3	0	-4	2	1
A_4	1	-5	3	0

Analyze the zero-sum game above. Determine the best strategy (mixed or pure) for each player and the expected payoff of the game.

6.5   Players A and B are playing a finger throwing game as follows. Each player simultaneously throws 1 or 2 fingers. Let s denote the total number of fingers thrown. If s is even, B pays A s dollars. If s is odd, A pays B s dollars.

a) Construct the game tree for this situation. You will need to identify one pair of nodes as an information set (why?).

b) Describe the possible strategies for each player and construct the normal form of the game.

c) Solve the game, obtaining the appropriate pure or mixed strategies and the final (expected) payoff.

6.6

	B_1	B_2
A_1	α	β
A_2	γ	δ

Consider the general 2 by 2 zero-sum normal form game pictured above.

a) Argue that, by relabeling strategies, it can be arranged that the $A_1 B_1$ payoff is less than or equal to all other payoffs. Conclude from this that there is no loss of generality in assuming that $\alpha \leqslant \beta$, $\alpha \leqslant \gamma$, and $\alpha \leqslant \delta$.

b) Assuming that payoff α is the smallest payoff (to A) as argued in a), show that the 2 by 2 game above has a saddle point and pure strategy solutions for each player *unless* $\alpha \leqslant \delta < \gamma \leqslant \beta$ or $\alpha \leqslant \delta < \beta \leqslant \gamma$.

c) Show that when $\alpha \leqslant \delta < \gamma \leqslant \beta$ or $\alpha \leqslant \delta < \beta \leqslant \gamma$, the game has the following mixed strategy solution:

$$\text{Player A: } (p, 1-p) = ((\gamma - \delta)/D, (\beta - \alpha)/D)$$
$$\text{Player B: } (q, 1-q) = ((\beta - \delta)/D, (\gamma - \alpha)/D)$$
$$\text{Expectation for player A} = (\beta\gamma - \delta\alpha)/D$$

where $D = \beta + \gamma - \delta - \alpha$. Why don't these formulas apply to the saddle point cases considered in b)?

d) Now drop the assumption that α is the smallest payoff. Then show that there are precisely six other orderings of α, β, γ, δ that yield mixed strategy solutions. Finally argue that the formulas derived in c) still give mixed strategies and payoffs for the games corresponding to these orderings.

6.7   A simple game is called *improper* if it has a pair of winning coalitions that have no members in common. Otherwise the game is called *proper*.

a) Let G be the simple game whose *minimal* winning coalitions are ABE, BFH, BC, CEFH, CDH. Show that G is an improper game, and discuss what might happen if players A, B, and E favored a certain action while players C, D, and H favored an incompatible alternative action.

b) By deleting a single MWC from the game G of part a), convert it into a proper game.

c) Prove that a general weighted voting game $[q; w_1, w_2, w_3, \ldots, w_n]$ is proper if $q > \frac{1}{2}(w_1 + w_2 + \cdots + w_n)$.

6.8   Recall the simple game SG of the text whose MWC's are AB, ACD, CDE. Prove that there is no choice of quota q and weights w_i that represents SG as a weighted voting game $[q; w_1, w_2, w_3, w_4, w_5]$. [Hint: assume such quota and weights do exist. Use the fact that AB is a MWC to conclude that $w_2 > w_3$. Show by analogous arguments that $w_1 > w_2$ and $w_3 > w_1$, thus establishing the desired result by contradiction.]

6.9   Consider the weighted voting game $[5; 3, 2, 1, 1, 1]$.

a) List all MWC's for this game. Call the players A, B, C, D, and E.

b) Compute the power of each of the 5 players in this game.

c) Compare the power of player B (having a weight of two) with the power of C, D, or E. Explain why this result is "paradoxical."

6.10   Methods for computing the Shapley-Shubik and Banzhaf power indices for a simple game are given below. Let n be the number of players in the game.

Shapley-Shubik

1) List all permutations (voting orders) of the players. There are $n!$ of them.

2) Call a particular ordering a *pivot* for player i if the coalition made up of all players *preceding* i in the ordering is not winning, but becomes winning when player i is added.

3) Define the

$$\text{Shapley-Shubik power of } i = \frac{\# \text{ of pivots for } i}{n!}.$$

Banzhaf

1) List all nonempty subsets (coalitions) of players. There are $2^n - 1$ of them.

2) Call a coalition containing i a *swing* for player i if the coalition is winning, but becomes nonwinning when i is removed. Note: a winning coalition may provide a swing for more than one player or no player at all.

3) Define the

$$\text{Banzhaf power of } i = S_i / S,$$

where $S_i = \#$ of swings for player i and $S = S_1 + S_2 + \cdots + S_n$, the total number of swings.

a) Compute Shapley-Shubik and Banzhaf power values for the game $BL^2 = (AC, AD, ACD)$. [Hint: the answers are $(2/3, 1/6, 1/6)$ and $(3/5, 1/5, 1/5)$.]

b) Argue for a general simple game that the individual powers must sum to 1 for each power index we have looked at: Deegan-Packel, Shapley-Shubik, and Banzhaf.

6.11 Apply the rules of Figure 7 to the HEXPAWN game tree of Figure 8 to complete the labeling of the nodes of the tree (with B's or W's). Then explain how it follows that Black (the player going second) will always win under perfect play. Finally, give a clear specification of a winning strategy for Black in Black's first two moves. Your strategy should, of course, anticipate all possible choices made by White in White's first two moves.

6.12 Consider a game between A and B whose payoffs are "win for A," "win for B," and "draw." Extend the reasoning of Figure 7 to draw three pictorial capture rules for a node at which it is A's move. Label the captured node with *captured by* B, *captured by* A, and *draw*.

## CHAPTER 7

Odds and Ends

Off to the races

There are, to be sure, many additional gambling and game-related topics with mathematically interesting foundations. In this chapter we consider a few of these. The choices we make will be somewhat arbitrary, based partly on popularity and the mathematics involved, but mostly on personal preference. We begin with the ubiquitous and moneyed activity of horse racing, "the sport of kings."

Horse racing has a rich and colorful history which we reluctantly bypass. We proceed directly to the elementary but interesting arithmetic underlying the pari-mutuel betting system used at all legal tracks (horse, harness, and dog) in the United States and Canada. Bettors may wager in amounts of $2 and up on a specific horse to *win* (finish first), *place* (finish either first or second), or *show* (finish among the top three). Amounts bet in each of these ways are registered by machines at the ticket windows, with separate *pools* for win, place, and show money. The ticket machines are linked electronically to the *totalizator* which computes and displays (on the *tote board*) pre-race totals for each pool, approximate current win odds for each horse, and post-race payoffs.

To illustrate the operation of the totalizator we run through the procedures for a fictitious five-horse race. The racing behavior of our horses will be governed by various well-defined randomizing techniques. This is clearly not realistic, but it will serve to review some of the probabilistic ideas of earlier chapters. We shall also see how the final payoff odds are determined solely by the betting that takes place, with the *morning line* (in our case true odds) only serving as pre-betting estimates made by some official "tout."

In Table 17 we list the horses, their finishing rules, and their "starting" order. The strange animals in our race run in the following manner.

TABLE 17

Horses, Post Positions and Rules for Finishing

Post Position	Horse	Requirement for Finishing
1	Double Six	Roll double sixes
2	Bi-Nomial	Exactly *3 heads* in 4 flips
3	Flip Ahead	Get a *head* in 1 flip
4	Roll Four	Roll a *four* on a single die
5	None Above	None of above horses finishes in given round

Finishing requirements are tested *in order of post position*. The first random event occurring defines the winner. After a winner is obtained (this must happen in the first "round" because of the finish rule for None Above), we start at the beginning of the list and repeat the randomizing activities for the remaining horses, continuing until a "success" occurs. This defines the place horse. The show horse is found by repeating this process a third time. (The probability that a place or show horse will never be determined is zero.)

The racing form for our strictly probabilistic horses is much simpler than a real racing form, provided the reader is well versed in elementary probability theory. Indeed, we can compute exact probabilities for all win, place, and show results for our race. Computations for place and show probabilities are quite messy, so we concentrate on the win probabilities and odds. This corresponds to the fact that racetracks only provide pre-race odds or a *morning line* for win bettors. The reasoning behind the win probabilities derives completely from the probability rules developed in Chapter 2. Thus,

$$p(\text{Win for Double Six}) = \frac{1}{36} \approx .028,$$

giving odds against of $(1 - \frac{1}{36})/\frac{1}{36} = 35$ or $35:1$.

The probability of getting exactly 3 heads in 4 flips of an honest coin is given by $C_{4,3}(1/2)^4 = 1/4$. Thus,

$$p(\text{Win for Bi-Nomial}) = p(\text{No Win for Double Six}) \cdot \frac{1}{4}$$

$$= \frac{35}{36} \cdot \frac{1}{4} = \frac{35}{144}$$

$$\approx .243.$$

Close and easily stated odds corresponding to .243 are 3 : 1 against. The remaining calculations proceed in this fashion until finally

$$p(\text{Win for None Above}) = \frac{35}{36} \cdot \frac{3}{4} \cdot \frac{1}{2} \cdot \frac{5}{6} = \frac{525}{1728} \approx .304$$

(about 5 : 2 against).

Table 18 gives the full theoretical story.

TABLE 18

Probabilities of Winning and "True" Odds Against

Post Position	Horse	Win Probability	"True" Odds Against
1	Double Six	.028	35 : 1
2	Bi-Nomial	.243	3 : 1
3	Flip Ahead	.365	3 : 2
4	Roll Four	.061	15 : 1
5	None Above	.304	5 : 2

Before or while the betting windows are open, the "rational" win bettor makes a personal estimate of the win probability for each horse. This estimate may be based upon many complex factors such as post position, past performance, jockey, track conditions, other horses, and inside information. Often other more mysterious factors such as colors, names, and superficial appearance of the horses can be brought into the "analysis." Bets are then made at some point during the fifteen minutes or so that the windows stay open before the start of the race. Table 19 gives some hypothetical win, place and show bet totals. The reader should note the 20 percent rakeoff (this figure varies from about 15

TABLE 19

Win, Place and Show Pools with Totals

Post Position	Horse	Win ($)	Place ($)	Show ($)
1	Double Six	975	1,000	600
2	Bi-Nomial	4,000	3,000	1,300
3	Flip Ahead	4,300	4,000	4,500
4	Roll Four	1,070	400	100
5	None Above	2,155	1,600	1,000
TOTALS		$12,500	$10,000	$ 7,500
−20 percent for track/state		− 2,500	− 2,000	− 1,500
Totals for payoffs		$10,000	$ 8,000	$ 6,000

percent to 20 percent in different locales) for the track and the state. This provides the "house edge" with a few small exceptions to be noted below.

Table 20 illustrates the computation for this win pool now made by the totalizator. As we shall see, the payoffs are determined completely by the overall totals for each horse in the win pool. We indicate how this is done by explaining the computations for Roll Four. As mentioned earlier, the morning line is estimated before the betting and is not, in the real horse world, a result of computation. To compute the payoff on a win for Roll Four, we take the $10,000 available for successful win bettors (Table 19) and divide it by the $1,070 wagered on Roll Four to win. This gives $10000/1070 = 9.346$ as the hypothetical dollar return for each dollar wagered on Roll Four. Since prices are based on the payoffs for a $2 ticket, this amount is doubled to 18.692. This figure is then rounded *down* to the nearest multiple of 10 cents to give a $2 bet payoff of $18.60 on a win for Roll Four. This figure, unlike the odds, is not posted unless the race is won by Roll Four. The last two columns show how much of the $10,000 win pool money is actually paid out, with the remainder or *breakage* providing added revenue to the track. The payoffs on the other horses are computed similarly. Note that the total payoff and breakage (which do not appear on the tote board) serve to increase slightly the already healthy house edge.

TABLE 20

Final Odds and Payoffs for Win Pool

Post Position	Morning Line	Amount to be divided: $10,000			
		Final Board Odds	$2 Bet Payoff	Total Payoff	Breakage
1	35 : 1	9 : 1	20.50	9,993.75	6.26
2	3 : 1	3 : 2	5.00	10,000.00	0.00
3	3 : 2	3 : 2	4.60	9,890.00	110.00
4	15 : 1	8 : 1	18.60	9,951.00	49.00
5	5 : 2	7 : 2	9.20	9,913.00	87.00

Computations for the place and show pools are more complex. They are not posted (and perhaps not performed) until the race has been completed. We present the calculations for a show pool based upon a 4-5-2 finish (Roll Four, then None Above, then Bi-Nomial) in Table 21.

The most subtle idea in Table 21 is the necessity of subtracting from the $6000 pot the total dollar amount of successful show bets ($2400 in our example) *before* dividing the pot by three. This subtraction must be done because successful bettors must first get their ticket purchase

TABLE 21

Payoffs for Show Pool when Three Top Finishers are #4, #5, and #2

Amount to be divided:		$6,000	
Amount of successful show bets:		$2,400 ($100 + $1,000 + $1,300)	
Amount left for division:		$3,600	
Divided 3 ways:		$1,200	
Post Position	$2 Bet Payoff	Total Payoff	Breakage
1 ----	0.00	0.00	0.00
2 Showed	3.80	1170.00	30.00
3 ----	0.00	0.00	0.00
4 Won	26.00	1200.00	0.00
5 Placed	4.40	1200.00	0.00
	TOTALS:	$3570.00	$30.00

payments back before actual profits are computed. Once this is done, the remaining pot is split three ways (one share for each of the successful show horses). Focusing on Roll Four again, we divide this $3600/3 = $1200 pot into 100 equal shares (from Table 19) to get a $12 profit for each $1 bet on Roll Four. This is then doubled to give a $24 profit on a $2 bet and hence a $26 payoff (the $2 wager is returned as well) for a successful $2 show bet on Roll Four. Payoffs on None Above and Bi-Nomial are obtained in similar fashion.

Several observations are now in order. The computations of Table 21 do not depend on the *order* in which horses 2, 4, and 5 finish as long as they are the first three across the line. On the other hand, if anyone of these horses finished "out of the money," *all* the payoffs would have to be recomputed (why?). Since order is unimportant, a full development of show payoffs would require $C_{5,3} = 10$ such tables! Another observation, which the reader may already have made, is that the show payoff on Roll Four well exceeds the win payoff. Table 22 shows the "official" payoffs on a 4-5-2 finish as they might be posted on the tote board. The "place" calculations are left for the reader.

TABLE 22

Final $2 Bet Payoffs on a 4-5-2 Finish

Post Position	Horse	Win	Place	Show
4	Roll Four	18.60	17.00	26.00
5	None Above	—	5.70	4.40
2	Bi-Nomial	—	—	3.80

The show payoff on Roll Four illustrates the idea known as a "hole in the show pool." Despite the fact that the show bettor has three ways to collect, the show payoff is better than the considerably less probable win payoff. The search for holes is an important strategic consideration for the sensitive bettor, so we elaborate upon it somewhat. The initial theory behind betting on holes is relatively easy. The bettor computes or estimates the percentage of money bet on a given horse in the place pool (the tote board provides current figures). This is repeated for the show pool. If either of these percentages is significantly less than the corresponding win pool percentage, the possibility of a hole exists. In addition to the computational challenge (the bettor does not have much time to do this), several factors complicate the process. Most seriously, lots of other bettors may also spot a glaring potential hole, and it is not uncommon to see your delicately selected hole fill in a hurry just before the race starts. Betting at the last minute may reduce this risk, but all the other hole seekers know this too. A second complication with hole betting is the fact that the payoff depends on which other horses split the pool with the one you select. If a favorite (or two) helps to fill out the place (or show) positions, the money to be divided will be greatly reduced after ticket prices are subtracted. The hopeful hole bettor's share is then reduced. If, on the other hand, longshots are involved, relatively less ticket purchase money will be subtracted (recall our show payoff on Roll Four) and the likelihood of a dramatic hole is greatly increased. This added uncertainty reduces the chances for startling payoffs such as the show payoff on Roll Four, but careful inspection for holes can still increase a bettor's expectation significantly.

This is another rare but intriguing phenomenon which can guarantee the bettor a positive expectation and cause the track to lose money on a race. A minimum return of \$1.05 is required by law on each dollar successfully bet (10 cents profit on each \$2.00 ticket). If betting on the favorite(s) is so heavy that the computed payoffs drop below \$2.10, the track must make up the difference. The following model shows one way this might happen.

Imagine a show pool where the combined dollar amount wagered on the three favorites is S, with the remaining dollars in the show pool summing to R. If all three favorites finish in the money, then we have

Show pool total:	$S + R$
$-$ 20 percent for track/state:	$- .2(S + R)$
Total for payoff:	$.8(S + R)$
Required payoff:	$1.05(S)$

The track will have to relinquish some of its "take" if $1.05S > .8(S + R)$ or $S/(S + R) > .8/1.05 = .762$. Thus the track will not be able to take its full 20 percent rakeoff if somewhat more than $3/4$ of the money in a pool was on horses that paid off in that pool. (The reader can check that our model can be applied just as well to the win or place pool.) Carrying this one step further, the track can actually lose money on a race if $1.05S > S + R$ or $S/(S + R) > 1/1.05 = .952$ (95 percent of the money on the favorite(s)). Situations where a favorite actually has a perceived .95 probability of winning are very rare. When they do arise, you can be sure there are some smart and heavy bettors who try to make the most of it. The track may have reduced profits on a day when this occurs.

As with other house edge gambling activities, track bettors sometimes win and occasionally win heavily. Unlike casino games (except blackjack), there are ways to obtain relative advantages over the "average" bettor. The mathematical approaches we have discussed can help, but far more valuable is extensive knowledge of the horses, their jockeys, and numerous "inside" factors. People with such knowledge can win fairly consistently at the track (though many more profess to such knowledge than actually have it). For the rest of us the virtually automatic 20 percent track edge becomes even higher, putting the beautiful sport of horse racing right alongside Keno as one of the least mathematically favorable of betting activities.

Lotteries and your expectation

An intriguing method of public fund raising is provided by an increasingly popular procedure known as the lottery. As governments seek additional ways of generating revenue, the state lottery often emerges as a creative way to extract voluntary contributions from a citizenry already sensitive to increasing taxes on sales, property, or income. While state lotteries are easier to implement and administer than race track betting, promoting and maintaining their popularity requires a blend of slick advertising methods and clever probabilistic design. In this section we take a close look at two specific games used in the Illinois State Lottery. Their description and design should provide illustrations of how typical state lottery tickets work. The mathematical analysis of these games will serve as further applications of the ideas on probability and expectation developed in earlier chapters. In addition, the analysis will show how to compute what individual ticket holders can mathematically expect and what state governments can count on from participation in a state lottery.

Figure 9 pictures the front and back of a ticket for POT OF GOLD, a relatively simple lottery game. As can be seen from the rules on the back, a 6 digit number will be randomly determined by means of a drawing. Payoffs will be given for tickets matching two or more consecutive digits, starting with the left hand digit and moving from left to right. Thus, if the 6 digit drawing number had been 058442 (a 3 digit left to right match), the ticket of Figure 9 would have been worth $50. If, on the other hand, the drawing number had been 158442, the ticket would be worthless since the initial left hand digit does not match.

To compute the expectation on this 50 ¢ ticket, we first determine the probabilities of matching exactly 2, 3, 4, 5, or 6 digits in order. Since there are precisely 1 million six digit numbers (000000 through 999999)

Figure 9   A typical POT OF GOLD lottery ticket.

each assumed to be drawn with equal probability, we have

$$p(6 \text{ digit match}) = \frac{1}{1,000,000} = .000001.$$

Similarly, 10 of these 1 million numbers will provide *at least* a 5 digit match. Since one of these is in fact a 6 digit match, we obtain

$$p(5 \text{ digit match}) = \frac{9}{1,000,000} = .000009.$$

TABLE 23

Probabilities and Payoffs for Pot of Gold

Event	Probability	Payoff	Expectation
2 digit match	.009	$ 5	$ .045
3 digit match	.0009	$ 50	$ .045
4 digit match	.00009	$ 250	$ .00225
5 digit match	.000009	$ 2,500	$ .0225
6 digit match	.000001	$90,000	$ .09[†]
TOTALS	.010000		$ .2250

Working backwards in similar fashion, we obtain the probabilities listed in the second column of Table 23. Note that these results correspond directly to information given on the back of the ticket as pictured in Figure 9. By summing them we see that a ticket holder has just one chance in a hundred of winning something on his ticket.

The payoffs corresponding to the number of digits matched are, except for the jackpot, specified directly on the back of the POT OF GOLD ticket.[†] Multiplying probabilities by payoffs for each event, we obtain the contributions to the expectation from each event in column 4. Summing these, we find that X(POT OF GOLD ticket) = $.225 = $22\frac{1}{2}$ cents. Since the cost of the ticket is 50 cents, it follows that the ticket purchaser can expect to lose $27\frac{1}{2}$ cents on each ticket purchased. The "state edge" on this game is, under our assumptions, a whopping 55 percent. Nowhere is it specified who is to be the recipient of this pot of gold.

[†] The fine print on the back of the ticket indicates that the payoff for a 6 digit match must be at least $15,000. A phone call to the Illinois State Lottery office determined that this jackpot payoff accumulates at the rate of $90,000 for each million tickets sold, so we use this figure for the payoff. In practice jackpot winners have always been determined before 2,500,000 tickets were sold.

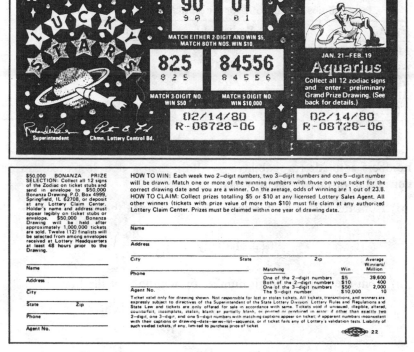

Figure 10   A typical LUCKY STARS lottery ticket.

The second game we analyze is described by the $1 LUCKY STARS ticket pictured in Figure 10. In this game a drawing is held to choose two 2 digit numbers, two 3 digit numbers, and one 5 digit number. Payoffs are awarded for matching one of the 2 digit numbers, both of the 2 digit numbers, one of the 3 digit numbers, or the 5 digit number. Further action, artfully tied in with current fascination with astrology, is available for dedicated LUCKY STAR players if they can collect all 12 zodiac signs.

Table 24 gives probabilities, payoffs, and expectations for the various matching events on a LUCKY STARS ticket. As we shall see, the computation of probabilities is more interesting and challenging than the analysis for POT OF GOLD. We first note that, since ticket and drawing numbers are randomly selected, duplicates (i.e. repetition of the same 2 or 3 digit number) on a single ticket or in the drawing can result.

To compute the probability of a double match (i.e., matching *both* 2 digit numbers on a ticket) we proceed as follows. When both ticket

TABLE 24

Probabilities and Payoffs for the Matching Events in Lucky Stars

Event	Probability	Payoff	Expectation
Match both 2 digit #'s	.000397	$ 10	$ .00397
Match one 2 digit #	.039006	$ 5	$ .19505
Match one 3 digit #	.001999	$ 50	$ .09995
Match the 5 digit #	.00001	$10,000	$ .10000
TOTALS			$ .39895

numbers coincide (probability $= 1/100$), then a double match can occur in one of three disjoint ways:

(i) from the first 2 digit number drawn, but not the second

$$\left(\text{probability} = \frac{1}{100} \cdot \frac{99}{100}\right);$$

(ii) from the second 2 digit number drawn, but not the first

$$\left(\text{probability} = \frac{99}{100} \cdot \frac{1}{100}\right);$$

(iii) from both the first and second numbers drawn

$$\left(\text{probability} = \frac{1}{100} \cdot \frac{1}{100}\right).$$

Summing the probabilities in (i), (ii), and (iii), multiplying by the $1/100$ probability of having identical 2 digit numbers on a ticket, we obtain

$$p(\text{double match with identical 2 digit ticket numbers}) = \frac{199}{1,000,000}.$$

For the case when the 2 digit ticket numbers differ (probability $= 99/100$), both of the drawn numbers must match the ticket numbers. Accordingly

$$p(\text{double match with different 2 digit ticket numbers})$$

$$= \frac{99}{100}\left[\frac{2}{100} \cdot \frac{1}{100}\right] = \frac{198}{1,000,000}.$$

Adding these probabilities, we conclude that the probability of matching both 2 digit numbers is .000397.

The probability of obtaining a single 2 digit number match is obtained in similar fashion. We summarize the reasoning in the expression below:

p(ticket #'s differ)

p(drawn #'s coincide)

p(drawn #'s differ)

$$\frac{99}{100}\left\{ \frac{1}{100}\cdot\frac{2}{100} + \frac{99}{100}\left[\frac{2}{100}\cdot\frac{98}{100} + \frac{98}{100}\cdot\frac{2}{100} \right]\right\}.$$

p(drawn # gives match)

p(left or right ticket # is matched, not both)

Performing the arithmetic gives a probability of $39,006/1,000,000 = .039006$ for a double match. We note this probability is mildly inconsistent with the "average winners/million" given on the back of the LUCKY STARS ticket. Computation of the remaining probabilities in Table 24 is more straightforward and we obtain expectations as shown in the right hand column.

Incorporation of the zodiac BONANZA part of the LUCKY STARS ticket requires more information than is given on the ticket and some simplifying assumptions. Since a determination of what percentage of a group of 1 million ticket holders will make themselves eligible for the zodiac subgame is impossible, we average things out as follows: The twelve BONANZA finalists, according to the *Illinois Lottery News*, receive payoffs with a distributions indicated in Table 25. Since these payoffs are awarded after 1 million tickets are sold, we can regard the zodiac BONANZA as offering twelve additional prizes, each with a probability of one millionth.

TABLE 25

Probabilities and Payoffs for the Zodiac Subgame in Lucky Stars

Event	Probability	Payoff	Expectation
Top Prize	.000001	$50,000	$ .05
Second Prize	.000001	$ 5,000	$ .005
Third Prize	.000001	$ 2,500	$ .0025
Fourth Prize (9 awarded)	.000009	$ 500	$ .0045
TOTALS	.000012		$ .062

From this table we see that the zodiac subgame contributes an expectation of 6.2 cents to the $1 LUCKY STARS ticket. Although the matching events on the ticket are not mutually exclusive, it still makes approximate sense to combine this figure with the expectations from Table 24 to obtain an overall expectation of 46.1 cents on the one dollar ticket. Thus LUCKY STARS provides a state edge of 53.9 percent, much the same as the state's edge in POT OF GOLD. Summing the individual probabilities in column 2 of Tables 24 and 25, we obtain an estimate (why would it not be exact?) that p(winning something) $\approx .0414$ or about 1 chance in 24.

As it turns out, all the different Illinois State Lottery games are designed to give the state an edge of about 55 cents on the dollar. This edge is in fact higher because of unclaimed prizes. During the revenue year ending June 30, 1978 almost 83.5 million dollars of Illinois lottery tickets were sold. Subtracting 33.7 million dollars for prize money and 12.7 million dollars for expenses and commissions, we arrive at a figure of 37.1 million dollars net revenue for that year. This profit, the state lottery publications are quick to point out, was distributed for human services, tax relief for the elderly, revenue sharing, and aid to educational institutions at all levels. From the point of view of state government, the lottery is apparently a successful and valuable institution.

Obviously the outlook is not so rosy for the individual citizen and potential lottery ticket purchaser. From a mathematical standpoint, the expectation of losing 55 cents on every dollar spent makes such lotteries by far the worst wager we have encountered in our tour of organized gambling activities. For some a dollar or so a week, even at such unfavorable odds, may seem a reasonable price for the excitement of following the lottery and dreaming about that BIG PAY DAY (the name of another Illinois lottery game). Others may spend more than they can really afford in their quest for action and instant wealth.

The gambler's ruin

In Chapter 5 we developed methods for determining the probability of being ahead (or behind) by more than a specified amount after a given number of bets in a fixed-odds game. By using the normal approximation to the binomial distribution and repeated computations, we were able to construct Table 15 to obtain these probabilities for a variety of situations. In this final section we consider the more absolute question of determining the probability that a gambler, with an unspecified number of repeated bets, will break the bank before going broke or vice versa. The answer we present to this problem, known as the *gambler's ruin*, has virtues of elegance and generality—just two tidy formulas will provide probabilities for all cases we might want to consider. In addition to supplying one more application of elementary mathematics to games and gambling, our results will enable us to take a look from another angle at a gambler's prospects against the house.

Suppose a gambler starts with i units of money and the casino with $T - i$ units, so that T represents the combined amount of money held by both the gambler and the house. The gambler plans to engage the house in repeated, 1 unit, even money bets in a fixed-odds game until either he or the house has all T units. Let p denote the probability that the gambler wins 1 unit on each independent bet, so that $q = 1 - p$ gives the probability that the house wins any given 1 unit bet.

Suppose that at some point in this sequence of "gamble to the death" bets the gambler has i units. From that point on the gambler's holdings may bounce around, rising or falling 1 unit with each bet, until his holdings reach T (we call this *success*) or 0 (we call this *ruin*). In either case, the game immediately stops since either the gambler or the house is broke. For each $i = 0, 1, 2, 3, \ldots, T$, let us define the crucial probability a_i by:

$$a_i = p(\text{the gambler succeeds given a current holding of } i \text{ units}).$$

Of course it then follows that

$$1 - a_i = p(\text{the gambler is ruined given a current holding of } i \text{ units}).$$

To warm up for the important observation that follows, we note that the gambler is broke if he has 0 units; therefore

$$a_0 = 0.$$

If his current holding is 1 unit, his only way to success is to win the next bet *and then* to succeed with 2 units. Hence

$$a_1 = p \cdot a_2.$$

With a holding of 2 units he has two ways to succeed: win the next bet and then succeed with 3 units *or* lose the next bet and then succeed with 1 unit. Thus

$$a_2 = p \cdot a_3 + q \cdot a_1.$$

When the gambler has all T units, he has succeeded:

$$a_T = 1.$$

Using the same reasoning that gave us the formula for a_2, we get the following set of equations among these $T + 1$ probabilities:

$$a_0 = 0, \qquad a_T = 1$$

(1) $$a_i = p \cdot a_{i+1} + q \cdot a_{i-1}, \quad i = 1, 2, 3, \ldots, T - 1.$$

The above set of equations is one example of an interesting and widely applicable mathematical construct known as a *difference equation*. While there exist various powerful and systematic techniques to solve such equations (i.e., to find an explicit formula for a_i, not involving the other a_j), we shall handle this one by elementary algebraic techniques. We work in steps, urging the reader to follow the algebra closely as we go along.

First we note that equation (1) above can be rewritten, using the fact that $a_i = pa_i + (1 - p)a_i = pa_i + qa_i$, as follows:

$$pa_i + qa_i = pa_{i+1} + qa_{i-1},$$

which is equivalent to

$$a_{i+1} - a_i = \frac{q}{p}(a_i - a_{i-1}).$$

Writing this out for each $i = 1, 2, \ldots, T - 1$, we obtain

(2)
$$
\begin{cases}
a_2 - a_1 = \dfrac{q}{p} a_1 & [\text{recall that } a_0 = 0] \\[2ex]
a_3 - a_2 = \dfrac{q}{p}(a_2 - a_1) = \left(\dfrac{q}{p}\right)^2 a_1 & \left[\text{since } a_2 - a_1 = \dfrac{q}{p}a_1\right] \\[2ex]
a_4 - a_3 = \dfrac{q}{p}(a_3 - a_2) = \left(\dfrac{q}{p}\right)^3 a_1 & \left[\text{since } a_3 - a_2 = \left(\dfrac{q}{p}\right)^2 a_1\right] \\[1ex]
\quad \vdots \qquad\quad \vdots \qquad\quad \vdots \\[1ex]
a_{i-1} - a_{i-2} = \dfrac{q}{p}(a_{i-2} - a_{i-3}) = \left(\dfrac{q}{p}\right)^{i-2} a_1 & [\text{reasoning as above}] \\[2ex]
a_i - a_{i-1} = \dfrac{q}{p}(a_{i-1} - a_{i-2}) = \left(\dfrac{q}{p}\right)^{i-1} a_1 \\[1ex]
\quad \vdots \qquad\quad \vdots \qquad\quad \vdots \\[1ex]
a_{T-1} - a_{T-2} = \dfrac{q}{p}(a_{T-2} - a_{T-3}) = \left(\dfrac{q}{p}\right)^{T-2} a_1 \\[2ex]
1 - a_{T-1} = \dfrac{q}{p}(a_{T-1} - a_{T-2}) = \left(\dfrac{q}{p}\right)^{T-1} a_1 & [\text{recall that } a_T = 1].
\end{cases}
$$

If we now add the left sides of all the above equations, something helpful happens—the sum collapses to $1 - a_1$ thanks to massive cancellation of intermediate terms. Adding the right hand sides and equating, we obtain

$$1 - a_1 = a_1 \left[\frac{q}{p} + \left(\frac{q}{p}\right)^2 + \left(\frac{q}{p}\right)^3 + \ldots + \left(\frac{q}{p}\right)^{T-1} \right].$$

Solving for a_1, we obtain

$$a_1 = \frac{1}{1 + \frac{q}{p} + \left(\frac{q}{p}\right)^2 + \ldots + \left(\frac{q}{p}\right)^{T-1}}.$$

Returning to the system of equations (2), we repeat the above procedure, this time working only with the first $i - 1$ equations in (2). We then obtain

$$a_i - a_1 = a_1 \left[\frac{q}{p} + \left(\frac{q}{p}\right)^2 + \ldots + \left(\frac{q}{p}\right)^{i-1} \right].$$

Solving for a_i and substituting the value for a_1 computed above, we finally obtain

$$(3) \qquad a_i = \frac{1 + \frac{q}{p} + \left(\frac{q}{p}\right)^2 + \ldots + \left(\frac{q}{p}\right)^{i-1}}{1 + \frac{q}{p} + \left(\frac{q}{p}\right)^2 + \ldots + \left(\frac{q}{p}\right)^{T-1}},$$

$$i = 1, 2, 3, \ldots, T - 1.$$

In these formulas for each a_i, the geometric progressions appearing in both numerator and denominator indicate that further simplification is possible. We break the analysis into 2 cases.

Case 1: $p = q = 1/2$ (a fair game)
In this case $p/q = 1$, and our results reduce to

$$a_i = \frac{i}{T}, \qquad i = 0, 1, 2, \ldots, T.$$

We thus are led to the simple and reasonable conclusion that, under our assumptions, the gambler's probability of breaking the bank equals the ratio of his current bankroll to the combined holdings of him and the house. Even if a bettor found himself magically engaged in a fair game with a casino, the relative insignificance of his holdings as compared to those of the house should strongly discourage any serious hopes of breaking the bank.

Case 2: $p \neq q$

In this case $q/p \neq 1$, and the geometric progression formula

$$1 + \frac{q}{p} + \left(\frac{q}{p}\right)^2 + \ldots + \left(\frac{q}{p}\right)^{k-1} = \frac{1 - \left(\frac{q}{p}\right)^k}{1 - \frac{q}{p}}$$

is applicable (the denominator on the right is nonzero). Applying this formula to (3) with $k = i$ and then $k = T$ and cancelling, we obtain

$$a_i = \frac{1 - \left(\frac{q}{p}\right)^i}{1 - \left(\frac{q}{p}\right)^T}, \qquad i = 0, 1, 2, \ldots, T.$$

To see the full impact of this general formula for the probability of breaking the bank, we first look at two special subcases.

Subcase 2a: $q > p$ and i is large enough that $(q/p)^i$ is "large" compared to 1. In this instance, the quotient $[1 - (q/p)^i]/[1 - (q/p)^T]$ is not significantly affected if the 1's are dropped in the numerator and denominator. Accordingly, we obtain $a_i \approx (p/q)^{T-i}$. Hence the probability of breaking the bank is determined by taking a probability ratio less than unity to a power equal to the number of units in the house's bankroll. We apply this result to an even money Las Vegas roulette situation ($p = 18/38$ and $q = 20/38$). Here $p/q = .9$, and even if the gambler and the house start with equal 50 unit bankrolls, (so that $(q/p)^i = 194.03 \gg 1$), the gambler's probability of breaking the bank is $(.9)^{100} = .005$. The gambler will be ruined 199 times out of 200.

Subcase 2b: $p > q$ and T is large enough that $(q/p)^T$ is essentially 0. In this positive gambler's edge situation, the denominator $1 - (q/p)^T \approx 1$. We thus obtain $a_i \approx 1 - (q/p)^i$ for the gambler's probability of breaking the bank, and $(q/p)^i$ for the probability of the gambler's ruin. In this unlikely setting, it is the house that faces almost sure ruin provided the gambler is given a reasonable starting bankroll (so that $(q/p)^i$ is close to 0) and plenty of time.

TABLE 26

Probabilities of Breaking the Bank on Even Money Bets

p = probability of winning on each independent bet
i = number of units the gambler starts with
T = total number of units held by the gambler and the house

	Optimal Blackjack	Fair Game	Craps	Roulette	Bookmaker
$T = 100$	$(p = .53)$	$(p = .5)$	$(p = .493)$	$(p = .474)$	$(p = .45)$
$i = 1$	.1132	.0100	.0018	.0000	.0000
$i = 5$	.4512	.0500	.0097	.0000	.0000
$i = 10$	.6992	.1000	.0209	.0001	.0000
$i = 25$	.9504	.2500	.0656	.0004	.0000
$i = 50$	.9975	.5000	.1978	.0055	.0000
$i = 75$	.9999	.7500	.4640	.0741	.0066
$i = 90$	1.0000	.9000	.7380	.3531	.1344
$i = 95$	1.0000	.9500	.8609	.5942	.3667
$i = 99$	1.0000	.9900	.9706	.9011	.8182
$T = 10,000$	$(p = .53)$	$(p = .5)$	$(p = .493)$	$(p = .474)$	$(p = .45)$
$i = 2$	.2136	.0002	.0000	.0000	.0000
$i = 5$	.4516	.0005	.0000	.0000	.0000
$i = 10$	.6992	.0010	.0000	.0000	.0000
$i = 25$	.9504	.0025	.0000	.0000	.0000
$i = 50$	.9975	.0050	.0000	.0000	.0000
$i = 9950$	1.0000	.9950	.2466	.0055	.0000
$i = 9975$	1.0000	.9975	.4966	.0741	.0066
$i = 9990$	1.0000	.9990	.7558	.3531	.1344
$i = 9995$	1.0000	.9995	.8694	.5942	.3667

In Table 26 we apply our formulas for a_i to compute probabilities of breaking the bank for various values of p, i, and T. The p values have been chosen to correspond to some of the gambling situations we have studied in this book. We assume all bets are even money (not quite correct for craps and blackjack) and that "optimal" blackjack play secures a 6 percent edge for the gambler.

The top half of the table considers situations where two adversaries have a combined bankroll of 100 units. While this figure is unrealistically low for a casino setting, the analysis of this section is, of course, applicable to interpersonal gambling situations as well. The values in the table show just how critical the value of p is for eventual success or ruin. The $i = 50$ row highlights this when starting bankrolls are equal. Clearly the relative sizes of i and $T - i$ are also important; but as p drops below .5, even larger values of i cannot withstand the probabilistic pressure of repeated play.

The lower half of Table 26 uses $T = 10,000$ in order to reflect circumstances closer to casino gambling (if each unit bet is $100, this would put the total bankroll at a million dollars—rather low for a casino). Here the importance of p is still more pronounced. Even uncharacteristically high values of i have little impact on avoiding the gambler's ruin where the edge is in favor of the house. The optimal blackjack column ($p = .53$) shows how dramatically things are reversed when the gambler manages to have a positive edge.

Table 26 yields no information about how many repeated plays might be required for the gambler's eventual ruin or success. Table 15 of Chapter 5 does offer some insight into the surprisingly large numerical answer to this question. More advanced mathematical techniques are available and provide a formal proof, but we simply assert that when p is close to $1/2$, the expected number of bets before ruin or success is approximately $i(T - i)$. If $i = 100$ and $T = 10,000$ this would give us 990,000 as the expected number of bets. Taking a rather optimistic figure of 500 bets per hour for 20 hours a day, this gives an expected gambling time of over 3 months before ruin or success. This example helps to illustrate why the positive edge held by a skilled blackjack player will not result in a quick cleanout of the casino. Working from the other end, we note that there are actually hundreds of gamblers playing simultaneously against the house. This leads to very much larger overall values of i, T, and $i(T - i)$ and helps to explain why casinos have not already accumulated all the gambling money available in the world—the expected time for ruin is large enough to allow new generations of gamblers to arise.

Answers to Selected Exercises

2.3   11 to 5 in favor of Walter

2.4 $p(C) = 206/216$; odds against B = 35 : 1

2.6 p(duplication with 4 dice) = $156/216 = 13/18$

2.7   Hint: compute probabilities of each of the following events: win both bets, win black and lose group of 4, lose black and win group of 4, lose both bets.

2.8   a) $X = -19/(37)^2$ dollars ≈ -1.39 cents

2.9 $X = 1/4$

3.1   a) $p = 3/4$;   b) $p = 25/36$

3.3 p(hit the blot) = $5/12$; p(enter unable to hit) = $1/3$

3.6 $p = 1/3$

3.10 "best" solution: $x = 3, y = 5$

4.2   a) 1260

4.9   c) $p = 363/C_{75, 5}$

5.1   b) $[1 - (\frac{5}{6})^6 - 5 \cdot (\frac{1}{6})(\frac{5}{6})^4] \cdot (\$10)$ (just over \$1.96)

5.3   b) approximately $1/300$

5.5   c) i) $p = .08$;   ii) $p = 0$

5.8   b) $X(\$1$ insurance bet) = \$2.00; X(no insurance) = \$2.10

5.9   a) house edge = -3.5%

6.4   A should play A_1 with probability $5/8$ and A_2 with probability $3/8$; X(player A) = $-11/8$ units

6.5   c) X(player A) = $-25/3$ cents

6.9   b) $p_A = 3/10$; $p_B = 3/20$; $P_C = P_D = P_E = 11/60$

Bibliography

References from the text

Bernoulli, J., *Ars Conjectandi*, Basel, 1713.

Cardano, G., *The Book of My Life*, translated by Jean Stoner, Dover, New York, 1962.

Cardano, G., *The Book on Games of Chance*, translated by Sydney Gould, in *Cardano, The Gambling Scholar* by Oystein Ore, Dover, New York, 1965.

David, F. N., *Games, Gods, and Gambling: The Origins and History of Probability and Statistical Ideas from the Earliest Times to the Newtonian Era*, Hafner, New York, 1962.

Dostoyevsky, F., *The Gambler*, translated by Jessie Coulson, Penguin, New York, 1966.

Huygens, C., *Calculating in Games of Chance*, from *Oeuvres Complètes*, 22 vols., M. Nyhoff, The Hague, 1888–1950.

Jacoby, O., *On Gambling*, Hart Associates, New York, 1974.

Jacoby, O. and Crawford, F., *The Backgammon Book*, Bantam, New York, 1973.

von Neumann, J. and Morgenstern, O., *Theory of Games and Economic Behavior*, Princeton University Press, Princeton, 1947.

Niven, Ivan, *Mathematics of Choice*, New Mathematical Library, M.A.A., Washington, 1975.

Ore, O., *Cardano, The Gambling Scholar*, Dover, New York, 1965.

Pascal, B., *Pensées*, translated by A. J. Krailsheimer, Penguin, New York, 1966.

Thorp, E., *Beat the Dealer: A Winning Strategy for the Game of 21*, revised edition, Random House, New York, 1966.

Other relevant works not referenced directly in the text

Banzhaf, J. F., "Weighted voting doesn't work: A mathematical analysis," *Rutgers Law Review* 19, Newark, 1965. (The original exposition of the Banzhaf power index.)

Berliner, H., "Computer backgammon," *Scientific American*, June 1980. (A sensitive account of the world champion's defeat by a computer robot.)

Deegan, J. and Packel, E., "A new index of power for simple *n*-person games," *International Journal of Game Theory* 7, 1978.

Epstein, R. A., *The Theory of Gambling and Statistical Logic*, Academic Press, New York, 1977. (An advanced mathematical treatment of some of the topics discussed in this text and many other gambling related topics).

Keeler, J. and Spencer, S., "Optimal doubling in backgammon", *Operations Research* 23, 1975.

Kemeny, J. G. and Snell, J. L., *Finite Markov Chains*, Springer-Verlag, New York, 1976. (A more mathematical and detailed treatment of the gambler's ruin problem.)

Rapoport, A., *Two-Person Game Theory*, University of Michigan Press, Ann Arbor, 1969.

Rapoport, A., *N-Person Game Theory*, University of Michigan Press, Ann Arbor, 1970.

Shames, L., "The Gospel according to X-22", *Esquire*, April 1980. (A sketch of the life and mind of Paul Magriel, one of the world's best backgammon players.)

Shapley, L. S. and Shubik, M., "A method for evaluating the distribution of power in a committee system," *American Political Science Review* 48, 1954. (The paper that introduced the first power index.)

Singleton, R. and Tyndall, W., *Games and Programs*: *Mathematics for Modeling*, Freeman, San Francisco, 1974. (An elementary exposition of the application of linear programming to two-person game theory.)

Steiner, G., "Fields of Force," *The New Yorker*, October 28, 1972. (An account of some recent developments in chess and connections with computers and artificial intelligence as of 1972.)

Selected references on specific games and activities

Magriel, P., *Backgammon*, Times Books, New York, 1976.

Phifer, K. G., *Track Talk*: *An Introduction to Thoroughbred Horse Racing*, Luce, Inc., Bethesda, 1978.

Reese, T., *Bridge for Bright Beginners*, Dover, New York, 1973.

Scarne, J., *Scarne's Guide to Modern Poker*, Simon and Schuster, New York, 1980.

Index